D0272903

The Quality of Frozen Foods

The Quality of Frozen Foods

Mogens Jul

Danish Meat Products Laboratory
Ministry of Agriculture
Copenhagen
Denmark

1984

ACADEMIC PRESS
(Harcourt Brace Jovanovich, Publishers)

London   Orlando
San Diego   New York   Austin
Montreal   Sydney   Tokyo   Toronto

TP
372.3
J84
1984

ACADEMIC PRESS INC. (LONDON) LTD
24/28 Oval Road
London NW1

United States Edition published by
ACADEMIC PRESS INC.
Orlando, Florida 32887

Copyright © 1984 by
ACADEMIC PRESS INC. (LONDON) LTD
Second printing 1985

All Rights Reserved
No part of this book may be reproduced in any form by photostat,
microfilm, or by any other means, without written permission
from the publishers

British Library Cataloguing in Publication Data

Jul, Mogens
The Quality of frozen foods.
1. Food, Frozen
I. Title
664'.02853   TP372.3

ISBN 0-12-391980-0

LCCCN 83-73403

Printed in Great Britain by
Whitstable Litho Limited

Contents

FOREWORD   xi

INTRODUCTION   1

History of food freezing   1

Quality aspects   3

Nutritional aspects   4

EARLY INVESTIGATIONS   5

Drip   5

Internal pressure   7

Freezing rate   7

Crystal formation   8

Freezing rate and quality   13

Contact plate freezing   18

Cases where freezing rate affects quality   19

Crystal growth mechanism   20

Crystal size and appearance   20

Persistence of small crystal theory   20

Irreversibility of water removal   22

Water migration 25

Importance of end temperature 26

Recrystalization 26

Temperature fluctuations and crystal growth 27

Freezing rate and drip 28

CONCLUSIONS REGARDING FREEZING RATES 33

Recommended freezing rates 33

Actual freezing rate 36

Equipment efficiency 37

Energy conservation and cost 38

Core temperature 40

QUALITY CHANGES DURING FREEZER STORAGE 44

Post freezing changes 44

Time-temperature-tolerance (TTT) 47

Triangle testing technique 50

Time of comparison 52

Stability and acceptability time 53

Acceptability factor 58

Scoring method for keeping quality determination 60

Reverse stability 64

Neutral stability 69

Determining shelf life   70

Accelerated tests   71

The logarithmic scale   73

Standardizing taste panel evaluations   76

Coordinating taste panels   76

Other factors than time and temperature   77

Comprehensive test   78

Objective tests   79

CHANGES IN NUTRITIVE VALUE   **81**

Effect of freezing on nutritive value   81

Meat's contribution to the nutritive value of diet   82

Meat intake in Denmark   86

B-vitamins in meat   86

Meat's contribution to B-vitamins in the Danish diet   90

Intake of frozen meat   90

Freezing meat for further processing   92

Institutional meat supply   94

Retention of B-vitamins in frozen meat   95

Influence of ageing   95

Influence of freezing   96

Rate of freezing   97

Influence of storage   97

Accumulated retention   98

Nutrients lost with drip   102

Vitamin B retention during kitchen preparation   104

Fluctuating temperatures   107

B-vitamin retention in frozen dishes   108

Comparing cook-freeze and warmholding procedure   108

Impact of meat freezing on the Danish diet   109

Effect on food habits   111

THE PPP-FACTORS   **112**

Product   114

Process   123

Packaging   128

Combined PPP-effects   135

Food legislation   136

Need for TTT-PPP data   137

ESTIMATED PRACTICAL KEEPING TIMES   **149**

Additivity of effect of various episodes   149

Exceptions to the rule of additivity   151

Calculating shelf life losses   153

COMPONENTS OF THE FREEZER CHAIN   **156**

The freezer chain   156

Time-temperature surveys   156

Residence times   165

Cold storage warehouse and wholesale storage   174

Transport   174

Product handling   180

Sales cabinets   182

Energy and retail cabinets   186

Improved cabinet management   188

Home storage   197

International trade   202

Temperature fluctuations   202

Temperature requirements   207

Temperature recommendations   208

ACTUAL SHELF LIFE CALCULATIONS   **209**

Data for the freezer chain   209

Probability of exposure   220

TTT-data   220

Actual estimates of end product quality   220

Limitations of data   221

Conclusions of calculations   244

Temperature indicators   245

Temperature abuse indicators   246

Temperature abuse integrators   247

Time-temperature integrators   247

Time-temperature recording   251

DATE LABELLING   **253**

Expiry date   253

"Use before" marking   253

"Best before" date   254

Packaging date   254

Maximum permitted keeping time regulations   255

ENERGY AND COST   **256**

Energy considerations   256

Commercial versus home freezing   259

THAWING   **261**

Effect on quality   261

Industrial thawing   263

Keeping quality of thawed products   265

Quality of thawed products   266

Labelling thawed products   268

Actual uses of thawed products   270

REFERENCES (bibliography, figures and tables)   271

INDEX   283

Foreword

Under the auspices of the intra-European collaborative research programme, COST, which has its Secretariat at the EEC Commission in Brussels, collaborative studies on various aspects of food technology have been initiated. One of these, the COST 91, concerns the influence of heat and cold in processing and distribution on quality and nutritive value of foods. One subject being studied is freezing and thawing. Research in this area is important, especially at the moment, because differing opinions exist in many areas of that field. Further, industry is continuously active in designing new products, improving packages, etc., and consumers are increasingly concerned with problems associated with the quality and nutritive value of their food supply. Furthermore, in Codex Alimentarius contexts and in many legislative or regulatory bodies, consideration is given to elaborating standards, rules or regulations regarding the production, distribution and handling of frozen foods. At the EEC Commission level, directives are also being considered. Additional data, which can be used in these contexts, are strongly indicated.

It is for this reason that under the COST 91 Project, a sub-group existed for collaborative work on the influence on quality and nutritive value of freezing, distribution and thawing foods, the Sub-Group 3. This group arranged a workshop, "Frozen Foods as Viewed by the Consumer", in Brussels, 8-9 December, 1980. The present work is a considerable elaboration of some data and views, which the author presented on that occasion. It is an effort to illustrate some of the uncertainties, and also some misconceptions, which exist in the area of frozen foods, and also to invite interested bodies and persons to give their opinions about the views expressed herein or to engage in further research into these and similar aspects.

It will be noted that this work contains several references to older literature and at times has omitted references to more recent and much more easily accessible data. This is done because it seems relevant to consider the historic development of our knowledge of the science of food freezing. It is the hope of the author that we may build our future on an intelligent interpretation and use of past experience and existing data. This in itself could, as illustrated in this work, at times be a departure from tradition, as regards food freezing and frozen food distribution.

In some cases, the author finds himself in disagreement with some widely held notions. This is not to be interpreted as a criticism of previous research and views. It is unlikely that we could have arrived at the present level of knowledge without the experiments and experience of the past, just as the views held in the present work will be subject to modifications as more data and insights become available, e.g. from the collaborative work sponsored by the above-mentioned Sub-Group 3 under the COST 91 Project.

It will be noted that this work is heavily biased with examples and references from meat and fish freezing, and much is based on Scandinavian findings; this is simply because the author has obtained most of his technical experience in these areas. Also, the author has over the years been involved in scientific and technical matters in relation to food freezing and thawing. This has mainly been as an adviser to industry, consumers or government, or as a teacher of students of food science or veterinary hygiene. Some of the opinions expressed have come from research and studies necessary for such activities. Much information has, however, also been received from persons active in various aspects of food freezing and the frozen food trade. Some of the views expressed are thus results of practical experience rather than scientific experimentation. Also, especially as a teacher, the author has often obtained information from text books, journals, conversations, lectures, etc., without always keeping exact records of the origin of the information. These facts will explain why references are sometimes not given and sometimes not complete.

The subject of this study merits further elucidation, because it seems that many conventional theories may be challenged. There is much to suggest that our present concepts of food freezing, and the distribution, storage, thawing, and use of frozen food in many cases are based on theories, which are characterized by being oversimplified explanations of experimental data. In some cases, data appear to have been overlooked or even misinterpreted.

Much of the information contained in this work was described in a book, "Industriel Levnedsmiddelkonservering 1-3" (Industrial Food Preservation), co-authored by the author with Erner Andersen and Hans Riemann, and published in 1966. An updated view on the same subject was published in the second edition of that work, entitled, "Konserveringsteknik 1-2" (Preservation Technology), co-authored by the author with Leif Boegh-Soerensen and Joergen Hoejmark Jensen, and published in 1978 (Vol. 1, revised 1983) and 1981 (Vol. 2). However, since these works both appeared in Danish, little of the information contained therein has received any attention outside of the Scandinavian countries.

In numerous cases, use has been made of tables, figures and data from "Food Technology", a publication of the US Institute of Food Technologists. The author is grateful for the kindness of the editor of this journal for permission to use this material freely.

The author wishes to draw attention to a series of articles, published in "La Revue Generale du Froid", July/August 1981, by Prof. R. Ulrich, President of the French National Council of Refrigeration. These articles were published under the title, "Variations de temperature et qualité des produits surgélés" (Ulrich, 1981). They cover a considerable number of subjects which are also covered in this work, albeit often from a somewhat different but certainly at least equally relevant perspective.

Some of the conclusions and findings described in the present work were subject to further scrutiny and research within the framework of COST 91. Better knowledge in these areas is likely to be the result and to be of benefit to both consumers and the trade and industry concerned once these data become generally available.

The present work was prepared as part of the author's contribution to the COST 91 Programme. Technical support in typing and the preparation of diagrams was received as part of a research grant from the Danish Council for Scientific and Industrial Research and from the Danish Council of Technology.

Credits are due to many who have helped, especially my very close collaborator in these matters, Leif Boegh-Soerensen, who advised on many matters and also prepared the index, to Ulla Boyter, who did the drawings, and to Lone Bro Petersen, who typed and almost perpetually retyped the manuscript.

Further, the author wishes to acknowledge the work of Henning Lüthje, who provided most of the data related to the effect of freezing on the nutritive value of meat, including most of the tables 30-56 and Fig. 46 and with whose consent the material is used in this work.

Hilleroed, December 1983   Mogens Jul

Introduction

History of food freezing

The use of freezing for food preservation dates back to pre-historic times. It was observed by primitive man that at continued low climatic temperatures perishable foods could keep almost indefinitely as long as they were maintained in the frozen state. Preservation was generally assumed to be satisfactory as long as the food was frozen hard, and little attention was paid to quality or nutritional value.

With the advent of mechanical refrigeration, it was an obvious step to construct rooms that would keep foods solidly frozen, i.e. at -5 to -7°C. Originally, no distinction was made between freezing rooms, i.e. rooms where the product's temperature is lowered, and freezer storage rooms, i.e. rooms where the low temperature of the frozen product is maintained. In the first years, the fact that products, mainly meat and fish, could be stored for considerable periods and then thawed and served as if they were fresh probably was such a source of satisfaction that not too much attention was paid to the quality loss with which the process unavoidably must have been connected. At any rate, the product would normally be found to be considerably superior to the dried or salted product, which otherwise might be the only available option, and thus a very welcome change from such foods.

Since then, freezing has become one of the more common methods of food preservation, both as a means of preserving raw materials for processing, preserving retail packaged food for sale to the ultimate consumer, and for the preservation of foods in the home where it, to a considerable extent, has replaced canning and curing. Finally, food freezing is used quite extensively in catering and institutional food supply systems.

As a means of food preservation, freezing is still rapidly increasing in use in many countries. Table 1 exemplifies this by giving some data for the consumption of frozen foods in France, quoted from Delaunay and Rosset (1981).

Table 1. Consumption of frozen foods in France, in tons, in 1978. After Delaunay and Rosset (1981).

	Surgélé (about −18°C)	Congélé (about −12°C)	Total
Vegetables	124 085	—	124 085
Potato products	44 734	107	44 841
Fruit and fruit juices	3 318	—	3 318
Meat	37 846	127 124	164 970
Poultry, rabbits	9 194	86 894	96 088
Fresh water fishes	4 825	7 503	12 328
Marine products	74 889	40 037	114 926
Bakery products	48 552	392	48 944
Ready-cooked products	19 794	1 074	20 868
Miscellaneous	575	1 175	1 750
Total	367 812	264 306	632 118

The same authors indicate that consumption of frozen products in France has increased with factors from 1.5 to 15 for various food groups in the period from 1973 to 1978. These authors also give the per capita consumption figures for frozen foods in various countries as listed in Table 2.

Table 2. Increase in per capita consumption of frozen foods in some countries, in kg, according to Delaunay and Rosset (1981), Dybfrostraadet (1982) and Dybfrostraadet (1983).

	1971	*1981*	*1982*
USA	34	40	
Sweden	11	20 (1979)	24.5
Denmark	11.5	22	23
UK	7*	15*	21*
Switzerland	5	11	16
FRG	4	8	16
France	3	8	10
Japan		5	
Italy	1	3	4

* Excludes poultry

Freezing foods in the home has steadily been gaining in importance as exemplified by Table 2 and Fig. 1, the latter showing the development in the number of home freezers in Denmark. Households with home freezers are fewer in some other countries, i.e. 63% in 1978 in the UK, according to Craig (1983).

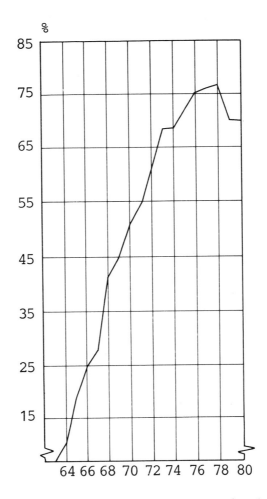

Fig.1. Per cent of households with provision for home freezing in Denmark. Dybfrostinstituttet (1981).

Quality aspects

Industry, trade, consumers, and science have, of course, long been interested in possible quality changes in foods due to freezing and subsequent thawing.

To begin with, most assumed that any change which was due to freezing and thawing would be a reduction in quality. This is, of course, not necessarily correct. Several changes during food preservation may not be noted by the consumer, and there are other cases where a measurable quality change will actually be perceived as an improvement; for instance, meat may often be considered somewhat more tender after freezing and thawing, cf. Table 3. Therefore, one needs to analyse each type of change separately and determine how it affects the quality of the food.

Nutritional aspects

Similarly, it is often assumed that any type of food processing or preservation, i.e. also food freezing, will result in a loss of a food's nutritional value and in an inferior diet.

This assumption is, of course, not justified, but the subject needs to be more thoroughly investigated. As will be illustrated in the following, freezing and thawing foods may often result in a loss of some nutrients, although the contents of others may, at times, actually increase. However, what is important, of course, is not the losses due to freezing, freezer storage and thawing *per se,* but rather the total losses, which occur from the time of harvesting, catching, slaughtering, etc., until the product is actually consumed. This loss is to be compared with the loss the same food undergoes when it is distributed fresh or in other forms and is prepared from that state by or for the consumer.

Further, equally important is the impact which the availability of frozen foods may have on people's diets and their nutritive value. Thus, frozen fish has to a large extent replaced salted and dried fish which in many cases led to increased fish consumption. The availability of frozen vegetables has often resulted in an increase in the consumption of vegetables, because they are available in seasons where they previously were either not available or available at a very high cost only. Because of changes such as these, freezing of foods may have a profound impact on nutrition in society.

Because of the increase in the use of frozen foods and because many questions still remain unanswered as regards the influence of this technique on quality, nutritive value and dietary changes, a closer examination of the influence of food freezing was found to merit inclusion of the subject in the COST 91 intra-European collaborative research programme.

Early Investigations

Drip

One of the phenomena which first came to the attention of industry, science and consumers was the fact that foods after freezing, freezer storage and thawing often exhibit a considerable amount of drip.

Drip may amount to 3-5% or even much more, see Table 3. In the early days of food freezing, this was often a sizeable financial loss, especially apparent in the wholesale trade, where most frozen products were then thawed. Also, frozen foods often appear somewhat dry and stringy, especially meat and fish. It seemed natural - and it was to some extent justified - to ascribe this quality characteristic to the drip with which it often seemed correlated.

Table 3. Tenderness and drip of beef, in 4 cm thick samples, aged 5 days before freezing, after freezing at various freezing rates and to various end temperatures. Shearing force determined in a Warner-Bratzler instrument. After Hiner, Madsen and Hankins (1945).

Freezing method		Decrease in shear force, % compared to unfrozen	Weight loss during thawing, %
18°F (-7.8°C)	still air	9.24	12.14
-10°F (-23.3°C)	still air	12.86	7.54
-40°F (-40°C)	still air	19.43	4.23
-40°F (-40°C)	air blast	23.85	3.75
-114°F (-81°C)	dry ice	28.55	1.91

It is shown below, however, that a direct correlation between drip and texture changes does not exist. Also, a certain drip loss may be compensated for by a correspondingly lower cooking loss. Therefore, a moderately high drip loss does not necessarily lead to a dry texture of a cooked product.

Drip has been ascribed to three factors, which are discussed in detail below, namely high internal pressures in the product, the effect of the formation of ice crystals in the tissue and irreversibility of water removal from the cells. All represented mechanisms that were easy to understand. Yet, in actual fact, they contribute relatively little to explaining the actual changes which take place.

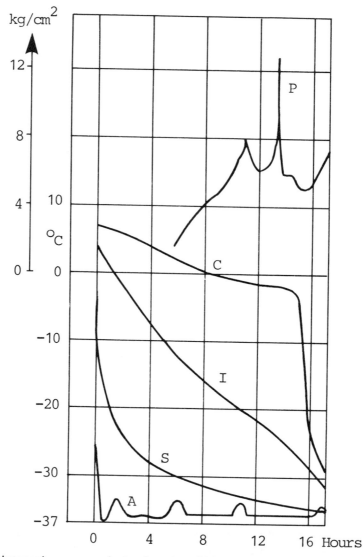

Fig. 2. Internal pressures during freezing of hind quarter of beef, at -35.5°C and 4.0 m/s air speed. The drops in pressure are due to physical ruptures because of pressure build-up. A) air temperature; C) temperature in centre of quarter; S) temperature at surface; I) temperature of intermediate layers; P) pressure in inner parts of quarter. After Lorentzen and Roesvik (1959).

Internal pressure

It has sometimes been assumed that freezing would cause actual ruptures in food products and that for instance drip was due to such ruptures. The theory was that the outer layers would freeze first and form a hard shell, which would resist the expansion of the inner parts when they eventually freeze. The result would be various ruptures. Measurements by Lorentzen and Roesvik (1959) resulted in the data in Fig. 2. It is seen that considerable pressures actually do build up when large pieces of meat are frozen. The sudden falls in pressure at various points do suggest that ruptures take place.

However, it seems that this factor is of little significance in practice as regards quality, and visual ruptures due to such pressures are few. In most cases, the tissue remains quite plastic for some time since water freezes out gradually, cf. Fig. 9. This will permit expansion to take place during freezing.

Freezing rate

A widespread view, which was accepted long ago and which still persists in some circles involved in producing, marketing and distributing frozen foods, among consumers, and occasionally even in technical and scientific circles, maintains that to obtain a satisfactory quality of frozen foods, these must be frozen at a very high freezing rate, "quick frozen", etc. Much experimental evidence is available in this area. The data suggest that apart from rarely used, very slow freezing, the rate of freezing has little, often practically no, influence on the quality of the frozen product. In addition, there are cases where very rapid freezing is detrimental to appearance or texture. Nevertheless, the above mentioned view, advocating quick freezing, is often expressed and frequently even incorporated in legislation, although the term is not always defined. Because of the popularity of this theory, it may be useful to discuss it at some length.

In so far as the author has been able to determine, the theory can be traced back to a work by Plank, Ehrenbaum and Reuter (1916). They carried out extensive experiments with the first rational freezing process which gained widespread acceptance, the Danish "Ottesen" method, which was patented in 1911. The introduction of the method into the USA is described by Clarence Birdseye (1951) who refers to it as the first quick freezing method in use in the USA.

To understand the situation, one must realize that in the early part of this century, foods were generally frozen simply by being placed in freezer storage rooms at about -7 to -10°C. No effort was made to separate the lots to be frozen from those already frozen. Moreover, boxes of fish, for instance, might be stacked together in such a storage room and left to freeze at the rate at which heat would be transmitted from the inner boxes. Thus, the process used as a control in Plank, Ehrenbaum and Reuter's experiments was still air freezing and storage at -7°C. Such freezing often resulted in a product of inferior quality. Contributing to this was the fact that freezing in those days was often resorted to only when sale in the fresh state for some reason had failed. This meant that the material to be frozen was often of somewhat questionable quality or might already be partly deteriorated. In addition, heat transmission from the inner parts of lots frozen that way was so slow that they might undergo microbiological or autolytic deterioration before they eventually froze and the temperature was reduced so much that microbiological activity terminated and enzymatic activity was no longer significant.

The virtue of the "Ottesen" method was that the product, primarily fish, was frozen in a brine at about -17°C. The brine was circulated at a very rapid rate around each fish, which meant that each piece was frozen within a few hours after being brought to freezing, so no microbiological or enzymatic deterioration could take place. Also, it is likely that since this was a relatively sophisticated method of freezing, high quality raw material was used.

Thus, the resulting high quality of the end product, believed to be due to the quick freezing, could well have been the result of a superior raw material and a very short interval between the products being placed in the freezer and the time freezing actually commenced.

One important characteristic of the "Ottesen" method, not directly relevant in this context, is that Ottesen had realized that by maintaining a brine concentration very close to the eutectic point, but with slightly less salt dissolved than at that point, very little salt is absorbed by the fish, thus avoiding any objectionable salty taste. Some salt absorption could not be avoided, a fact that did not impair taste directly, but it catalyzed some rancidity which gave problems under extended storage. It was actually primarily this factor which eventually led to the method being abandoned some 40-50 years later.

It was because of the German patent office's demand that this specific characteristic of the Ottesen freezing method be proven before a patent was granted, that the process was extensively investigated by Plank, Ehrenbaum and Reuter (1916).

Crystal formation

Considering that one of the characteristics of food freezing is that water freezes out and forms ice crystals, they carried out extensive histological studies on the effect of the Ottesen freezing method on the structure of tissue compared with what was then conventional freezing.

Reuter demonstrated, cf. Fig. 3 and 4 that very fast freezing rates, e.g. freezing small pieces of tissue in liquid nitrogen or carbon dioxide snow, resulted in the formation of small ice crystals within the cell structures, some ice formation in intercellular spaces, and comparatively little change of the histological picture of microtomic preparations of the tissue.

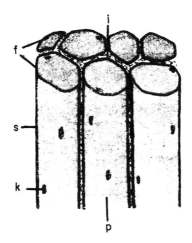

Fig. 3. Diagram of unfrozen fish tissue. f) Muscle fibres; s) Sarcolemma; k) Nucleus; p) Plasma; i) Intercellular collagen. (Plank, Ehrenbaum and Reuter, 1916).

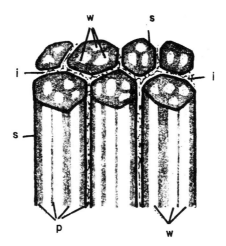

Fig. 4. Diagram as Fig. 3. but muscle frozen in small pieces in liquid carbon dioxide. w) indicates ice crystals. Freezing time 5-10 min. (Plank, Ehrenbaum and Reuter, 1916).

If fish were frozen in brine, he found in the tissue immediately under the surface, the formation of several crystals inside the muscle fibres, but still relatively little change in the histological appearance, cf. Fig. 5.

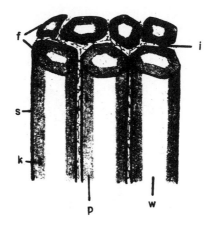

Fig. 5. Diagram as Fig. 4 of the structures of rapidly frozen fish tissue, taken directly under the skin of brine frozen fish. (Plank, Ehrenbaum and Reuter, 1916).

If, however, freezing was somewhat slower, he found that comparatively large ice crystals were formed, partly inside the cell walls and seemingly causing ruptures of these, and partly between the cells, cf. Fig. 6.

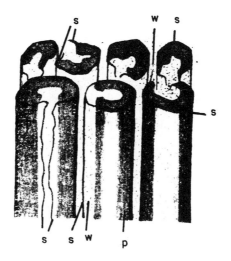

Fig. 6. Diagram as Fig. 5 but from deeper layers of brine frozen fish. (Plank, Ehrenbaum and Reuter, 1916).

Very slow freezing rates in still air at -7°C resulted in crystal formation in intercellular spaces only, cf. Fig. 7.

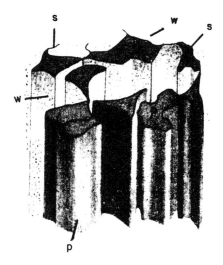

Fig. 7. Diagram as Fig. 5 but from air frozen fish. Since most of the water was removed from the cells, such tissue appeared highly distorted when examined under the microscope. (Plank, Ehrenbaum and Reuter, 1916).

These histological investigations were not carried out at actual freezing temperatures. Instead, the frozen tissue was fixed in the frozen state, sliced and dyed, and then examined in the conventional way. Holes observed in the tissue were - correctly -assumed to be indicators of size, shape and location of ice crystals. The justification of this interpretation of the histological studies was later confirmed by Woolridge and Bartlett (1942a and b). Very recent studies of frozen tissue by Bevilacqua and Zaritzky (1980) and Calvelo (1981) show essentially the same picture as have many similar studies over the years.

Nusbaum *et al.* (1983) give some similarly obtained pictures and data which are particularly interesting because they are carried out on ground tissue, i.e. ground beef, cf. also Table 8.

The early studies by Reuter demonstrated that very rapid freezing with intracellular ice crystal formation would cause the least damage to the organization of the cell tissue in the frozen state. It was concluded, somewhat summarily, that this would result in the least change of taste and texture due to freezing.

This explanation had another attraction in that in those days freezing was mainly carried out for wholesale purposes. The products were often thawed and sold as fresh. As mentioned above, one very annoying thing was that thawing was accompanied by a certain amount of drip. For slowly frozen beef, this might amount to 5% or higher, cf. Table 3, for fish even more, and thus constitute a very substantial economic loss. When one considers the considerable disorganization of the cell structure after slow freezing shown in

Fig. 7, it seemed obvious that fast freezing would result in less drip than slow freezing.

A closer examination of these experimental data, however, makes the conclusions thus drawn from these studies less convincing when offered as explanations for the need to freeze food quickly in order to obtain good quality. Thus, Fig. 4 shows, compared to the unfrozen fish muscle in Fig. 3, that very fast freezing does, indeed, result in relatively little physical change in the tissue. However, Fig. 4 relates to fish frozen in small pieces in liquid carbon dioxide with extremely rapid freezing rates and is not illustrative of any commercial food freezing process in use in those days. Fig. 5 also shows frozen tissue with limited physical disorganization of the tissue. However, this sample was taken immediately under the skin of a brine frozen fish and is thus not typical of the main parts of the product. Conversely, Fig. 6 shows tissue from the inner part of a brine frozen fish and is illustrative of the cell structure changes in the major part of quickly frozen fish. It shows extensive cell wall rupture and might well be expected to result in a higher drip than slow air freezing, as shown in Fig. 7 where at least no cell wall ruptures are observed as ice crystals seem to have been formed in intercellular spaces only. If, upon thawing, this water would diffuse back into the cells - as is actually the case to a large extent - one might actually expect less drip as a result of such freezing. Thus, the theory advanced was actually based on a misinterpretation of the data on which it was based, because it could only explain the effect of such rapid freezing rates which were not in use in those days, and, actually, can only be achieved where food in small pieces, e.g. peas, shrimps, pellets or drops, are frozen almost instantaneously in very cold liquids, e.g. brines, etc., or directly in the freezer medium, e.g. freon, carbon dioxide or liquid nitrogen.

In reality, this meant that the quick freezing, considered desirable and defined as one with the formation of small crystals, mainly intracellularly, and with very little cell wall rupture, was not obtained and not obtainable with the freezing process recommended except in insignificant amounts of surface layers. Therefore, the virtue of the process could not be ascribed to small crystals or intracellular crystal formation as was actually done. The widespread acceptance of this theory is, however, also due to misinterpretation of the conclusions of these authors because Plank, Ehrenbaum and Reuter (1916) clearly state that the main changes in a frozen product occur not during the freezing process but during freezer storage.

In addition, in the conclusions regarding quick freezing versus slow freezing it was arbitrarily assumed that the least visual change in appearance of the frozen tissue would represent the least quality change. Reuter did demonstrate that the change in appearance due to slow freezing was retained to some extent, also in the thawed tissue. However, no attempts were made to determine the relationship between these tissue differences and organoleptic quality of the ready-to-eat product. Later data, e.g. those of Nusbaum *et al.*

(1983) fail to show any justification for drawing any conclusions between microstructure in the frozen state and palatability of the thawed tissue.

However, for years, the theory of quick freezing leading to small and intracellular crystal formation and, therefore, to superior quality of the frozen product, was rarely questioned.

Freezing rate and quality

In 1932 Moran (1933) tested freezing beef fillets, sheep's liver and loin of lamb. Some samples were frozen rapidly, i.e. within 50 minutes; others were frozen slowly, i.e. in 15-20 hours. The samples were tested organoleptically both immediately after freezing and after up to 4 months' storage at -10 or -20°C.

Table 4. Evaluation of the frozen sides from six experiments on unwrapped pork side. After Clemmensen and Zeuthen (1974).

Exp. No.	Initial product temp. °C	Air temp. °C		Air velocity m/s		Freezing time h.	Freezing loss %	Visual evaluation of samples
		Initial	Final	Initial	After the surface temp. has reached -6°C			
1	7	-25	-31	2.7	unchanged	3.33	0.8-1.3	5-10% of exposed meat surface discoloured
2	11	-25	-31	5.7	unchanged	2.60	0.75-1.2	50% of exposed meat surface discoloured
3	10	-25	-30	1.0	unchanged	4.50	0.9-1.3	5% of exposed meat surface discoloured
4	12	-25	-30	1.0	2.7	4.00	0.9-1.35	5-10% of exposed meat surface discoloured
5	8	-30	-37	1.0	2.7	3.70	0.6-1.3	5% of exposed meat surface discoloured
6	7	-20	-27	1.0	2.7	5.00	0.8-1.6	5-10% of exposed meat surface discoloured

The test results were quite definite: there was no discernible difference between the organoleptic characteristics of the slowly and the rapidly frozen products for the same period of storage.

Conversely, data such as those given in Table. 4 indicate how surface desiccation may influence product quality for an unwrapped product and that too rapid air blast freezing may actually be deleterious to quality. Zaritzky, Anón and Calvelo (1982) indicate how liver gets an undesirable whitish appearance if frozen rapidly.

Thus, the relation between freezing rate and quality is not at all clear. Prestamo (1979) reports an increased tendency for developing freezer burns in slowly frozen green beans; the effect is ascribed to relatively faster water evaporation from the larger ice crystals in the tissue.

In investigations of meat microstructure, Carroll, Cavanaugh and Rorer (1981) find more damage to the microstructure in very fast, i.e. cryogenic freezing, compared to samples frozen at -18°C for 24 hours, i.e. quite slow freezing. They conclude, however, that eating quality does not appear to be adversely or positively affected by very fast freezing.

Experiments on freezing electrically stimulated lamb in New Zealand as described by Chrystall, Hagyard, Gilbert and Devine (1982) showed the freezing time to a temperature of -4°C in the deep round should be more than 12 hours after entry into the freezer. Otherwise, the meat will be tough. The Meat Industry Research Institute of New Zealand even developed a monitoring device to enable freezer operators to verify that this freezing rate was not exceeded.

Many experiments on the influence of freezing rate have been carried out. Some results are summarized in Tables 5, 6, 117 and Fig. 8.

Table 5. The organoleptic qualities of beef steaks after various freezing methods (-5 = extremely poor, +5 = extremely good). After Buchter and Zeuthen (1970).

	Texture	Taste	Juiciness
Blast –40°C	3.35	2.87	3.57
Plate –40°C	2.99	2.83	2.85
Liquid nitrogen	3.05	2.99	2.89

Most tests failed to show any significant difference between moderate and fast freezing rates. Any advantage of obtaining very small crystals was rarely established, which caused Heiss (1942) to state that the crystal formation theories and the conclusions drawn with regard to quality of frozen foods had been based mainly on a simplistic interpretation of physical phenomena which seemed exceedingly plausible but, unfortunately, had turned out to be unable to explain differences in quality. Yet, as indicated above, the theory persists to this very day. Especially advertising and trade promotion and even much

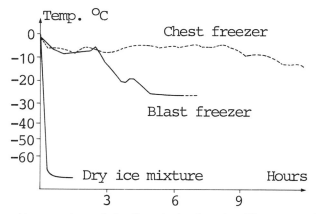

Fig. 8. Internal temperatures in broilers during freezing. These samples were all taste tested. In spite of the very great differences in freezing time, no taste difference was detected in the tests. After Miller and May (1965).

Table 6 . Effects of various freezing methods on organoleptic characteristics of different fishery products. LFF = liquid freon freezing. (Score: 1 = poor, 5 = best). After Aurell, Dagbjartsson and Salomonsdottir (1976).

Product/Factor	LFF	Blast	Plate	Block
Shrimp				
Odour	4.4	4.4	4.4	4.4
Flavour	4.1	4.1	4.4	4.1
Texture	4.3	4.3	4.3	4.3
Scallops				
Odour	3.8	3.9	3.9	—
Flavour	3.8	3.7	4.0	—
Texture	3.8	3.9	3.9	—
Lemon sole				
Odour	4.5	4.1	4.3	—
Flavour	4.5	4.3	4.1	—
Texture	4.2	4.0	4.2	—

teaching in food science and home economics seem to contribute to its longevity. Even Sebranek (1982) gives support to the theory.

In general, many hold the view that fruits and vegetables benefit from rapid freezing due to the cellulosic composition of the cell walls.

Freezing rate is probably somewhat more important in the case of vegetables and fruits. Thus, Morris and Barker (1933) reported considerably better results with freezing fruits very quickly than by slow freezing. However, after a short storage period at -20°C, the difference disappeared.

A very comprehensive study carried out by Heiss (1942a and b) did not confirm this view. Only where freezing times were very long did texture seem to suffer. Conversely, taste generally was optimal at moderate freezing, at times superior to that at both very short and very long freezing times. He concluded that in so far as organoleptic quality was concerned there would be no advantage in using freezing rates higher than what was then generally in use in industry, i.e. in the case of Germany, freezing packaged products in ordinary blast freezers.

An advantage is found for freezing fruits in that their form is better preserved after quick rather than slow freezing, cf. Table 7 which quotes results by Stoll *et al.* (1977).

Table 7. Organoleptic quality of various varieties of string beans, after quick and slow freezing. After Stoll, Dätwyler, Fausch and Neidhardt (1977). (Score: 1 = poor to 10 = excellent).

	Shape	Taste	Texture
Quick frozen (3 cm/h)	8.8	8.4	8.5
Frozen slowly (still air 21°C)	7.4	8.4	8.1

Lenartowicz *et al.* (1979) found little effect of freezing method on flavour and colour of frozen fruit, but some effect on shape and drip.

The advantages of quick freezing and a somewhat improved texture for fruits and vegetables seems by and large to be observed for products which are eaten uncooked, simply because cooking results in extentive denaturation of much the same type as that which is caused by freezing and thus masks any difference between a slowly and a quickly frozen product. Especially for potatoes, however, some benefit seems to be derived from fast freezing even where the product is cooked.

The author made a review of some of these data in 1944 (Jul, 1944/45). The cell wall rupture theory offered a mechanical explanation which was easy to understand. It failed to take into account biochemical factors, e.g. water binding characteristics. In meat or fish technology it is well-known that ground tissue has a considerably better water binding capacity than the unaltered tissue. This is due to the release of myofibrillar proteins which serve as effective water binders. Grinding the tissue is similar in many ways to rupturing cell walls by ice crystals. Thus, cell wall rupture *per se* might well result in improved water binding and decreased drip rather than the opposite. That this mechanism actually exists was demonstrated by Hiner, Madsen and

Hankins (1945), who found more cell rupture by freezing at and to -40°C than at and to -l8°C. In this case, an increase in cell wall rupture actually reduced drip; this was assumed to be due to the above mentioned release of water soluble proteins as in the case of ground meat.

As discussed later, drip following the thawing of frozen products is to a large extent due to protein denaturation during freezer storage, which is particularly pronounced at warm frozen storage temperature, i.e. -5 to -8°C as was common in the early days of food freezing.

While the explanation originally offered for the superiority of the Ottesen freezing method is open to considerable questions, there was no doubt that in those days a great improvement was made by freezing foods relatively rapidly compared to the methods previously used. This should be ascribed to the fact that quick freezing came to mean freezing of superior quality raw material, frozen in individual pieces, e.g. cuts of meat, individual fish, etc., rather than freezing stacks of boxes placed in cold storage rooms often without ample air circulation around the product, at temperatures not colder than -10°C. Under the latter conditions, products were exposed to ultraslow freezing and often underwent some microbiological or enzymatic deterioration prior to freezing.

In a recent experiment, where ground beef patties, (hamburgers), were frozen unprotected in liquid carbon dioxide (freezing rate 5 cm/h), air blast at -30°C (1 cm/h), air blast at -15°C (0.4 cm/h), and still air (0.3 cm/h). Nusbaum *et al.* obtained the results summarized in Table 8.

Table 8. Weight loss and tenderness score of unpackaged, individually frozen hamburgers. (Score: extremely tender = 9, extremely tough = 1. After Nusbaum *et al.* (1983).

Freezing rate cm/h	5	1	0.4	0.3
Weight loss during freezing, %	0.5	1.5	2.8	4.5
Cooking loss, %	18.4	19.0	21.3	24.1
Tenderness score	6.7	6.5	6.0	5.3

It is seen that in reality, little is achieved in tenderness by increasing freezing rate from 0.4 cm/h to 5 cm/h. In interpreting this, however, one should be aware of the considerable difference in total preparation loss, 18.9% for very rapid freezing to 28.6% for very slow freezing. In the author's experience, such differences would more than fully explain the differences in tenderness scores. The experiment seems to confirm that no disadvantage in palatability is achieved by very rapid versus quite slow freezing. In stating this, it must be mentioned that in this experiment, the samples must have been frozen to very different end temperatures, presumably varying from approximately -80 to -10°C, a fact which may be of greater importance for palatability than the differences in freezing rate, cf. section: Importance of end temperature below.

From a practical point of view, the experiment may show that economically, it may be an advantage to freeze quickly, but hardly in liquid carbon dioxide even if it in some cases gives lower shrinkage during freezing if small objects have to be frozen unpackaged. On the other hand, it must be realized that regardless of effects on palatability, freezing unprotected small units of products in air at -10 or -15°C is not likely ever to be considered even in primitive home freezing because of surface desiccation.

Contact plate freezing

The theory of the need for very quick freezing got a further boost when Clarence Birdseye introduced his contact plate freezer in the late 1920s. Here again a superior quality was obtained. It was assumed to be due to the high freezing rates obtained when products were frozen in quite low packages pressed between hollow freezer plates and at low evaporator temperatures, about -40°C.

As an interesting and somewhat revealing comment, the author may describe how he in the late 1930s had discussions with Birds Eye, Inc., personnel in the USA and questioned the necessity of the use of the plate freezer. The company officially maintained that this was very important for product quality. However, the author then remarked that he had seen large quantities of Birds Eye's labelled products being sharp frozen at Seabrooks Farms, Inc., a large vegetable packer in New Jersey, supplying frozen vegetables under the Birds Eye label. It was then explained that the Birds Eye company had done much to find varieties of fruits and vegetables suitable for freezing. Further, much work had gone into developing suitable packaging; in those days, the company even maintained control over the freezer cabinets in retail outlets. It was then conceded that the superior quality of the Birds Eye brand products was due to these factors, not to any superiority of the freezing method as such. However, such details were not suited for use in advertising and trade promotion because any packer might claim that he took similar, quality enhancing steps. If, on the other hand, the superior quality of the Birds Eye products in all trade promotion was ascribed to the Birds Eye patented process, no other supplier could match the claim.

The plate freezer has one feature which later came to be a much appreciated quality attribute in that retail cartons are frozen pressed between plates. Therefore, they attain a completely flat, smooth surface, which makes the package quite attractive. In other types of freezing retail cartons, expansion of the contents may cause the package to become somewhat uneven and untidy.

Cases where freezing rate affects quality

Especially in the early days of freezing, much interest was shown in cases where live tissue may be frozen and thawed and still remain alive. It is thought that when freezing, and maintenance in the frozen state, and thawing is possible without the organism being severely damaged in a physiological sense, very little change could have taken place with regard to taste.

Such cases are numerous also in the vegetable world, although it is known that most plants are protected against freezing or at least against excessive ice formation in their tissue in that they build up high concentrations in the fluid phase of the tissue, i.e. lower the freezing point. Thus, many trees will not survive sudden, heavy frost, but the gradual coming on of winter leads to a building up of dissolved matter in the cell liquids and results in protection against frost.

In the animal world, such cases are known mainly from the lower classes of animals which survive very cold winters.

In the higher animal kingdom, cases are also known where survival is possible even after prolonged cold storage, namely in the freezing of semen and embryos. These are frozen very quickly, generally in liquid nitrogen, and it must be assumed that no ice crystal formation takes place at all, but that water forms amorphous ice. It is characteristic that they can only be successfully stored at very low temperature, e.g. in liquid nitrogen. If storage temperature is increased to anything like that used in commercial cold storage of frozen foods, ice formation starts to take place, and survival is no longer possible.

Some similar cases where very rapid freezing appears advantageous may be mentioned. Freezing tomato slices very rapidly, e.g. in liquid nitrogen, may result in very small ice crystals inside the cells or even amorphous ice formation. Such tomatoes will retain a suitable texture when thawed. However, this applies for only a brief period or for storage at very cold temperatures, because otherwise recrystalization takes place with subsequent crystal growth, cancelling out any beneficial effect on texture.

Sebranek (1982) quotes several cases where cryogenic, i.e. very rapid freezing of fish and shellfish leads to better texture, an advantage said to be retained during freezer storage. Cryogenic freezing is necessary for freezing some dairy products, boiled eggs etc.

Conversely, very rapid freezing has at times led to quality deterioration. For instance, where fish fillets or cuts of meat are frozen in liquid nitrogen, the surface layers are often frozen so rapidly that considerable internal strain develops, surface ruptures occur and an inferior product is the result, cf. Rasmussen and Olson (1972).

Carroll, Cavanaugh and Rorer (1981) called attention to the somewhat more extensive cell structure disorganization which may be caused by very rapid, e.g. cryogenic freezing.

There are cases where very large ice crystals have been formed and have resulted in inferior product quality. Cases have been recorded where ice crystals with dimensions of up to 50-70 mm have been observed. They had resulted in actual cuts in the tissue and caused an unsightly appearance of the product. In all cases they seem to have been due to recrystalization during freezer storage at relatively warm freezer temperatures and not to too slow freezing. They were the results of actual temperature abuse compared to what would be considered reasonable storage practice even very long ago, i.e. temperatures warmer than -8°C.

A slight loss in contents of B vitamins, due to freezing per se was reported by Lehrer *et al.* (1954) for pork.

Crystal growth mechanism

It has been questioned if the theory of crystal growth and a resulting cell wall rupture is not too mechanistic for another reason. Crystals do not grow like for instance plants but rather by the adherence of water molecules to already formed crystals. Thus, these will grow only where free water molecules are present. However, studies by Woolridge and Bartlett (1942a and b) using a crystallographic technique and examining the samples in the frozen state show that ice crystals actually grow and penetrate cell walls and thus puncture them. Studies by Bergh (1949) on the actual physical growth of crystals during freezing confirmed this view.

Crystal size and appearance

The above has referred mostly to taste and texture. One should keep in mind that other quality factors also apply. Thus, in the case of some poultry, especially turkeys, very rapid freezing is normally used in the initial stages of freezing. Frequently, turkeys are wrapped in plastic bags and crust frozen in a glycogen solution. They are subsequently air frozen in conventional air blast freezer tunnels. The initial rapid freezing results in very fine ice crystals in the surface layers and gives the product a very whitish appearance which is preferred by the trade and most consumers.

Persistence of small crystal theory

The theory about the correlations between quality, small ice crystals, and cell wall rupture has remained widely accepted in spite of the many data which indicate that very rapid freezing rarely leads to particularly high quality of the frozen product; in addition it is only seldom achieved. As already mentioned, it seems that the theory is longlived because it is easily understood. It may also be supported by equipment manufacturers who may obtain by sophisticated - and expensive - freezing installations fast rates of freezing. The sale of such equipment is advocated by suggesting that it will result in superior products, statements which are rarely supported by experimental data; probably, manufacturers are themselves convinced about the validity of their claim. Finally, the frozen food trade has often seen a sales advantage in reminding its customers that the products have been quick frozen and therefore can be assumed to be of superior quality. As illustrated above in the example from the Birds Eye organization, an easily understood explanation for superior quality is a very valuable asset for promotional purposes regardless of its validity. The theory is given considerable support and prominence in a popular publication about frozen foods published by the Belgian association of frozen food companies (CRIOC, 1982).

Because of the persistence of the theory, many food laws, regulations,

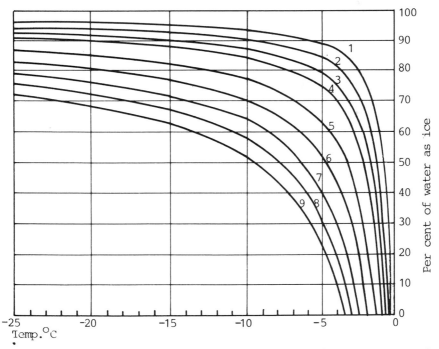

Fig. 9. Percentage of total water frozen out as ice in various foods at various end temperatures. 1) Egg; 2) Milk, fish (hypotonic); 3) Meat, tomatoes; 4) Onions, apples; 5) Beans, carrots; 6) Apples, pears, plums, potatoes, isotonic fish (sharks and rays); 7) Citrus fruit; 8) Grapes, cherries; 9) Bananas. After Heiss (1936).

standards or codes of practice require deep frozen foods to be frozen very quickly, within a specified time limit, etc., see section: Core Temperature below. In the UK it was, for a period, required that herrings should be frozen so that temperatures should be reduced from 0°C to -5°C in two hours or less. It is probably factors such as these that caused Mascheroni, Anón and Calvelo (1981) to suggest a laboratory test to determine, by examination of frozen beef, the rate with which the product has been frozen.

A demonstration of the fact that the freezing rate is normally of negligible importance for quality is seen in the fact that most people are perfectly satisfied with the quality of home frozen foods, also when these are compared to commercially frozen products, cf. for instance Fig. 8. Home freezers are hardly ever so equipped that freezing rates obtained therein could be designated as anything but moderate to slow.

Irreversibility of water removal

When organic tissue is frozen, ice formation takes place. The dissolved substances in the remaining liquid become more concentrated and its freezing point drops. Figure 9 illustrates this situation for various foods. In thawing, the process is reversed. However, one might ask if the water once removed from its position as partly or fully bound to proteins, carbohydrates, etc., will actually return completely to its position in the system or will remain at least partly as "free" water, easily lost in thawing, i.e. drip.

Table 9. Freezing points and ice formation for some foods. Data from Worksheets published by the Deutscher Kältetechnischer Verein.

Product	Water content %	Freezing point °C	Ice in % of total water °C					Ice in % of product °C				
			-5	-10	-15	-20	-30	-5	-10	-15	-20	-30
Lean meat	74	-1	74	82	85	87	88	55	61	63	64	65
Cod	80	-1	77	84	87	89	91	62	67	70	71	73
Egg, yellow	50	-2	80	85	86	87	87	40	43	43	43	43
Egg, white	86	-1	87	91	93	94	94	75	78	80	81	81
Peas	76	-0.9	64	80	86	89	92	49	61	65	68	70
Spinach	90	-0.7	88	93	95	96	97	79	84	86	86	87
Raspberries	83	-1.2	87	91	94	96	97	72	76	78	80	81
Fruit juice	88	-1	72	85	90	93	96	63	75	79	82	84
White bread	35	-4	5	33	43	45	46	2	11	15	16	16

Table 9 and Fig. 9 show the situation for various foods. Since many foods are salt cured, and some fruits are frozen with sugar, corresponding data for some salt and some sugar solutions are given in Table 10.

Table 10. Freezing points and ice formation in some pure solutions. Data from Worksheets published by the Deutscher Kältetechnischer Verein.

	Sugar/Salt content %	Freezing point °C	Ice in % of total water and ice °C					Ice in % of product °C				
			-5	-10	-15	-20	-30	-5	-10	-15	-20	-30
Sugar	10	-0.7	92	96	100	100	100	83	88	90	90	90
Sugar	25	-2.0	55	75	82	100	100	41	56	62	75	75
Sugar	35	-3.5	27	60	70	100	100	18	39	42	65	65
Salt	2	-1.3	77	88	91	93	100	75	86	89	91	98
Salt	3	-1.9	66	81	87	88	100	64	79	84	85	97
Salt	7.5	-5	7	50	65	70	100	6	46	60	65	93
Salt	15	-11	0	0	25	34	100	0	0	21	29	85
Salt	20	-16.5	0	0	0	14	100	0	0	0	11	80
Salt	25	-21	0	0	0	0	100	0	0	0	0	75

Table 11. Salt concentration (salt-in-brine), i.e. $NaCl \div (H_2O + NaCl)$ in some cured meats.

	%
Liver paste	2
Cooked ham	3
Sliced bacon	7
Fermented sausage	15
Dry sausage	20
Dry sausage	25

Table 11 shows the percentage of salt in the liquid phase of some cured meats. When fruits are frozen with sugar, the sugar content is generally about 25% and the sugar concentration in the liquid phase about 34%. The data in Table 10 can, of course, not be used directly when estimating the effect of freezing on sweetened or salted foods since the liquid phase in these foods will contain other dissolved components than sugar or salt respectively. Thus, the percentage of water frozen out will be lower in actual foods than might be expected from the data in Table 10.

On the basis of this, a theory was developed by Rudolf Plank, Germany, that the end temperature in the freezing process should not be too cold. It was argued that at very cold temperatures, even the "chemically" bound water would be removed from the molecules and the reversibility of water back to its position in the tissue would be jeopardized. In this it was assumed that the quality of a frozen food could be measured by the degree to which the changes brought about by freezing could be reversed by thawing, thus the "reversibility" of the process seemed important. This theory was illustrated as reproduced in Fig. 10, which suggests that a cold freezing temperature will improve quality because it results in the formation of small ice crystals but also that freezing should be terminated before the tissue reaches too cold a temperature because of the assumed irreversibility of water removal in the form of ice crystals.

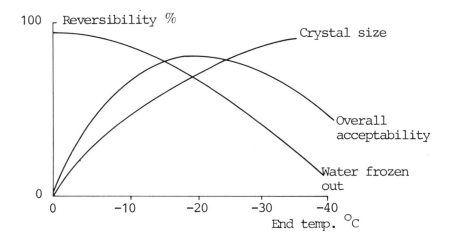

Fig. 10. The concept of the reversibility of freezing. Each curve indicates that factor's contribution to reversibility. After Tuchschneid (1936).

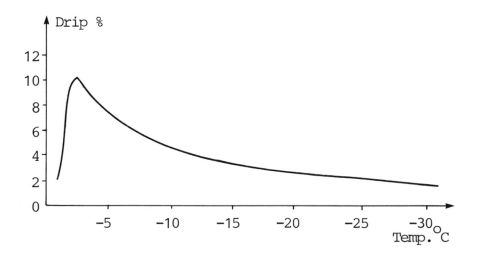

Fig. 11. Drip according to end temperature in beef. After Andersen, Jul and Riemann (1966).

A closer consideration will show that this, indeed, was an oversimplification. Figure 9 shows that there will normally be less than a few per cent difference in the amount of water frozen out at say -20°C and at a much colder temperature. Since drip may exceed this figure several times as seen in Table 3, and since irreversibility was considered closely related to drip, the theory was easily shown to be incorrect. Thus, Fig. 11 shows how drip may be no higher when freezing to a cold end temperature is used, see also Table 3.

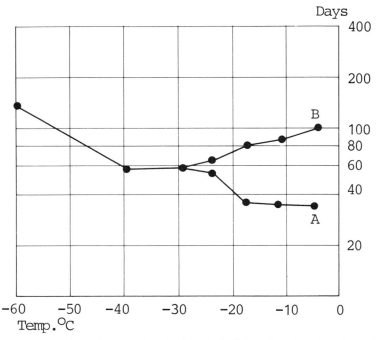

Fig. 12. Shelf life of nitrite cured salted pork bellies, sliced and packaged in oxygen permeable film. A) Samples frozen to an end temperature of -40°C and stored at temperature indicated. B) Samples frozen at the temperature at which they were stored. After Lindeloev (1978).

Water migration

Because of the lack of uniformity of temperature in a tissue during freezing, cf. Figs 20 and 21, there will, of course, be different concentrations in the liquid phase of a product during freezing. This may result in a certain diffusion due to differences in osmotic pressure, a process which may only be partly reversed during thawing. Drip may actually be associated also with this effect. Also, when, as described later, moderate to slow thawing is often advocated, the explanation given is that of allowing for more time for equalization of the concentration of dissolved matter in the various sections of the product.

Importance of end temperature

While cold end temperatures do not lead to increased drip, some dis-advantages may occur as the result of low end temperatures. A few cases have been observed where other, irreversible damages are observed at low end temperatures; thus, certain fruits discolour when frozen to a very low temperature. Also, the practice of freezing fruits in a sugar solution or covered with sugar may to some degree be related to some factor of this kind. Adding high concentrations of sugar to the fruits will, of course, result in a water phase with a high sugar content, i.e. a low freezing point. Thus, less water will be frozen out in the case of a sugar frozen product than in a non-sugared product at the same end temperature.

Interesting data regarding this aspect were obtained by Lindeloev (1978) as illustrated in Fig. 12. Freezing loosely packaged bacon to a cold end temperature, -40°C, gave shorter keeping times at -5 and -12°C than freezing the product to the storage temperature only. It is possible that some irreversible water removal takes place at the cold end temperature, affecting subsequent shelf life at comparatively warm storage temperatures.

Similarly, Winger (1982) found that lamb frozen to an end temperature of -5°C gave a considerably longer shelf life at an accelerated test with storage at -5°C than lamb frozen to an end temperature of -35°C.

Recrystalization

Tied to the consideration of size of ice crystals obtained in freezing has often been speculation with regard to and studies of the possibility of crystal growth during freezer storage through recrystalization. It is well-known that in many crystalline products one may originally have small crystals. However, the smallest are unstable compared to the larger ones. Under certain conditions, therefore, there will be a growth in crystal size, cf. Reid (1983). One might thus assume that even where small ice crystals were obtained by very rapid freezing, crystal growth during storage would take place and obscure the picture, e.g. cause cell wall rupture. It is well-known that such growth will be greatly accelerated by fluctuating temperatures.

Figure 13 shows some measurements where a growth in crystal size was recorded. While such phenomena are known to occur, they have not been found causative for quality problems in frozen products, except in cases where fluctuating temperatures warmer than -12°C are encountered. In such cases one may observe, as noted by Strachan (1983), mass movement of water within the product to such a degree that direct damage occurs.

The theory behind recrystallization phenomena is discussed by Calvelo (1982).

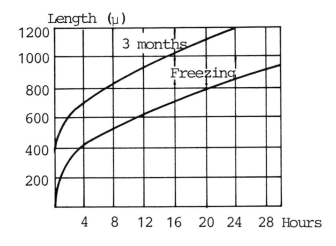

Fig. 13. Average length of ice crystals in frozen tissue after various freezing times and after 3 months' storage at -20°C. After Persson (1980).

Moran (1932) reported significant crystal growth at 6 months at -3°C, none at -20°C. Both Fennema, Powrie and Marth (1973) and Bevilacqua and Zaritzky (1982) found crystal growth at warm temperatures, but not below about -10°C.

When frozen breads are kept at relatively warm freezer storage temperatures, so-called white rings may develop. These are due to migration of moisture and recrystallization, cf. Pence (1969).

The tenderizing effect of freezing foods, especially meat, sometimes reported, may be ascribed to either formation of large crystals because of slow freezing or because of storage at warm temperatures, e.g. - 14 to -5°C.

Temperature fluctuations and crystal growth

Although more related to storage phenomena, the fear of growth of the small crystals obtained by very quick freezing is probably the reason why on seemingly purely theoretical considerations, it is generally recommended that freezer storage rooms be maintained at constant temperatures and that all temperature fluctuations be avoided. As is indicated later, this fear has been vastly overrated. In few cases is ice recrystalization observed at temperatures colder than -12°C. Further, it is rare that fluctuations in freezer storage rooms would amount to more than ±2°C with fairly infrequent fluctuations. No adverse effect on quality has been recorded under these conditions which could be traced back to such fluctuations.

Experiments concerning the influence of fluctuating temperatures are reviewed by Ulrich (1981). Little evidence of any reduction in keeping quality due to temperature fluctuations was disclosed at temperatures below -18°C.

However, when any beneficial effect of small ice crystals is claimed, it is probably important to study the effect of temperature fluctuations, as it might apply to frozen foods stored in retail sales cabinets. As is indicated below, the temperature of packages in retail cabinets may vary from -8 to -23°C. Further, highly fluctuating temperatures are normally found in sales cabinets. Also, the fact that consumers and store personnel often move the packages around either for inspection or to give space for new stock results in considerable temperature fluctuations because products are frequently shifted from one temperature zone to another. One example of this is the very significant in-package desiccation which is commonly observed in sales cabinets for most products which are packed loosely in transparent plastic bags, cf. Table 98. These temperature fluctuations may also have an adverse effect on product quality, but the author is aware of only few studies which specifically have covered this temperature range and fluctuations of such nature which can occur in retail cabinets, cf. Tables 79 and 85 and Figs 87 and 94.

Freezing rate and drip

As already indicated, cf. Table 3, one of the quality factors which at times show some relation to freezing rate is drip. Fig. 14, where the thawing method

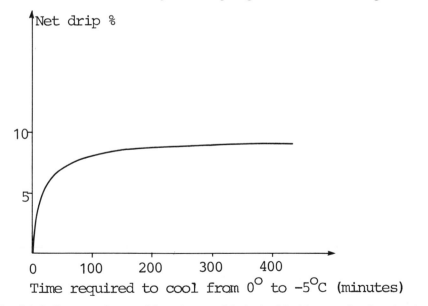

Fig. 14. Influence of rate of freezing on drip in haddock's muscle, thawing in running water at 8 to 10°C. After Reay (1933).

was standardized, indicates that drip was lower after very rapid freeezing. However, in the normal range of freezing rates used commercially, it is practically independent of freezing rate.

Similar results were found in experiments with cuts of 4-5 year old beef cattle, where Cook, Love, Vickery, and Young (1926) found the relationship shown in Table 12, cf. also Table 3.

Table 12. Relation between freezing rate and drip for beef, according to Cook, Love, Vickery and Young (1926). Reprinted by permission from CSIRO.

Freezing rate cm/h	Drip %
1,000	0
2.8	1.4
2.3	1.5
1.6	3.4
1.0	3.9

On the other hand, Empey (1933) failed to establish any correlation between freezing rates and drip. In some cases, an increase in drip has been reported as a result of cryogenic freezing.

Drip seems to depend somewhat less on freezing method than is often expected. Thus, a Dutch method has been worked out, whereby the amount of extraneous water, which has been taken up in chicken during the slaughtering process, including water chilling, may be determined by measuring the amount of drip which the bird gives off when thawed under well-defined conditions. This method has been extensively tested and has been authorized for use in determining water uptake for official control in the EEC, as stipulated in EEC Regulation 2967/76.

Randi Pedersen (1982) found for water chilled chicken broilers that drip amounted to from 52 to 72% of the water taken up. While the method is found only moderately reliable, a reasonable correlation between actual water pick-up and drip does exist. This may confirm that drip is not much affected by minor differences in freezing procedures but possibly also that the freezing procedures used in the European poultry industry do not vary a great deal. However, this appears to be an aspect worthy of further study; some rather surprising data obtained by Crigler and Dawson (1968) are given in Fig. 15. These authors worked with a wide range of freezing rates for poultry and found that the drip varied quite inconsistently with freezing rate.

While the effect of thawing on drip is not very great, data vary between different experiments. Thus, Vail *et al.* (1943) and Westerman *et al.* (1949) found that thawing rate had little effect on drip. When animal tissue is frozen pre-rigor, slow thawing gives less drip.

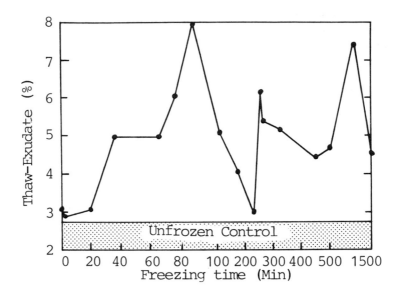

Fig. 15. Effect of freezing time on the percentage of drip from chicken breast muscles. After Crigler and Dawson (1968).

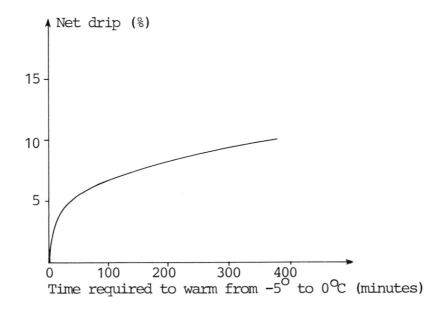

Fig. 16. Influence of rate of thawing on drip in haddock's muscle, brine frozen at -21 °C. After Reay (1933).

Tanaka and Tanaka (1956) found much lower drip when large pieces of whale meat were thawed by dielectric heating than by slow thawing in air.

Figure 16, where freezing methods have been kept constant, indicates that thawing methods may influence the total amount of drip. It suggests that for fish slow thawing rates result in the highest amounts of drip.

Conversely, Calvelo (1981) found decreasing drip as thawing time was increased for beef.

Morris and Barker (1933) provided the data on drip from strawberries as related to freezing rate presented in Table 13. Limited effect of freezing rate on drip was found.

Table 13. The effect of freezing rate on drip from strawberries. After Morris and Barker (1933).

Method of freezing	Time from +5° to –5°C	Drip %
Liquid air	1 min.	53.5
Brine	29 min.	59
Air –20°C	6.3 hrs.	62.5
Air –10°C	10 hrs.	61.3

A detailed literature study of freezing rate as related to drip and quality was given by Kondrup and Boldt (1960).

The author (1969) reviewed various experiments regarding the influence of freezing rate on drip in meat; it was concluded that very differing results are obtained in different experiments.

Honikel and Fischer (1980) have demonstrated for meat that in cases where thaw rigor may occur, speed of thawing can have a pronounced influence on drip. Thus, beef muscle was found to have about 7% drip after slow thawing and storage at 0°C for 4 days but up to 20% drip after rapid thawing and storage at 0°C for the same period. Some of the apparent inconsistencies in the above mentioned results may be due to unreported interference from such factors.

Ristić and Tawfic (1975) found drip increasing from approximately 1% for chicken broilers immediately after freezing to 5.5% after 24 months storage in the frozen state at -20°C.

Anón and Calvelo (1980) found for beef frozen in sample sizes of about 100 grammes under rigorously controlled conditions an exudate of about 14 to 19% at very fast freezing, increasing to 21 to 22% at a freezing time of 17 min. and then again a smaller drip, almost independent of freezing rate, at longer freezing times. They ascribe this to the assumption that up till a freezing time of 17 min., crystal formation is mainly intracellular, above that it is assumed to be mainly extracellular. (The very high exudate percentages found are due to the sample actually having been cut in slices and centrifuged).

Because of the high variation between freezing times in various depths of beef being frozen, these results appear difficult to interpret. They have not been replicated by others.

Conclusions Regarding Freezing Rates

Recommended freezing rates

Possibly because of the somewhat ambiguous views prevailing about the effect on quality of freezing rate, limited attention has often been paid in industry to actual freezing rates. Thus, large lots of fruits or meat are sometimes frozen in large cartons in palletized stacks in freezer tunnels. Air spaces or rather air channels permitting air circulation in between cartons are often not provided. Then, in spite of the rapid air blast and the low temperature, which is often assumed to result in a high freezing rate of the product, the very mass of the material may be such that the inner parts undergo very slow freezing and may

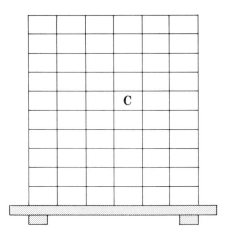

Fig. 17. Cartons with pre-cooked dishes, placed in outer cartons and then palletized and frozen in blast freezer. In the inner packages, at C, microbiological deterioration actually took place before freezer temperatures were reached. After Löndahl and Nilsson (1978).

therefore often deteriorate microbiologically or enzymatically before they are frozen, Fig. 17 shows such a case.

Workers in New Zealand have reported the growth of yeasts in the inner parts of bulk frozen meat, due to continued high temperatures in these parts due to too large block sizes during the first part of the freezing process. Among practical freezer operators, there is often incomplete comprehension of such factors as air temperature, evaporator temperature, air velocity, the need to separate cartons during freezing, actual temperatures in the product, etc.

To meet such conditions, some very practical plastic divider plates have been developed. They appear somewhat like the well-known egg-trays. They are convenient, also in that they take up very little space when not in use because they stack easily.

On the basis of practical experience and literature surveys, Andersen, Jul and Riemann (1966) arrived at some conclusions with regard to recommended freezing rates for various food products which form the main basis for the following suggestions:

Beef. Beef does not seem to be affected by variations in freezing rate between 0.2 and 5 cm/h. A high freezing rate may cause the meat to be slightly lighter in colour both before and after thawing, cf. also Zaritzky, Anón and Calvelo (1982). No effect of freezing rate on contents of B vitamins were found, e.g. by Lee *et al.* (1954).

For cuts of beef a relation exists between freezing rates and drip. In one case a freezing rate of about 1.5 cm/h gave a drip of 2 per cent, while under comparable conditions a freezing rate of 0.25 cm/h gave a drip of 8 per cent. It must be noted that drip is also highly dependent on the size of the cut, the method of cutting, and probably also on the thawing procedure.

Poultry. The taste of poultry does not seem to be affected by variations in freezing rates from 0.15 cm/h to 5 cm/h. The surface of frozen poultry is lighter when the freezing rate for that is above 1.0 cm/h. Therefore, as mentioned above, turkeys particularly are often crust frozen at a high freezing rate and then subjected to normal air blast freezing.

Also for poultry, drip seems somewhat dependent on freezing rate. In one experiment, drip varied from 1 at a freezing rate of over 1 cm/h to 2-3 per cent at a rate of 0.1 cm/h.

Fish. No difference in taste was found between freezing rates of 0.15 cm/h and 5 cm/h for cod fishes. Normally, a freezing rate of not less than 0.5 cm/h is recommended. Any increase in freezing rate over that does not seem to bring about any improvement. In some cases, slowly frozen fish tissue was reported as more tender than a quickly frozen product. In other cases the taste of the slowly frozen product was judged to be more like that of the unfrozen product than the taste of a quickly frozen sample. The texture of fish is often reported to be better after rather slow freezing rates.

For fish the percentage of drip is generally higher at low freezing rate.

Very high freezing rates, e.g. 20 cm/h, have given an unattractive appearance of the frozen product, and have had an adverse effect on taste. It seems that this quality loss may be due to the freezing rate itself and not only to the low end temperature. When the temperature of a normally frozen fish was subsequently reduced to -139°C, a much smaller effect of the low temperature on quality was found, compared to freezing the fish directly at -180°C and thus very rapidly. Some shellfish, e.g. individually frozen shrimp and mussels, seem to benefit from fairly rapid freezing, e.g. over 1 cm/h.

Fruits and vegetables. The taste of fruits and vegetables is approximately independent of those freezing rates that occur in normal practice. It does seem, however, that immersion freezing of small pieces of vegetables, e.g. peas, results in a quality improvement. The freezing rate may then be approximately 20 cm/h.

The texture of vegetables and fruits suffers somewhat from freezing because of changes in the cellulosic cell walls. This change seem to be somewhat affected by the freezing rate, e.g. freezing rates for asparagus under 0.5 cm/h may result in a somewhat stringy texture, cf. also Table 117.

Table 14. Quality characteristics of strawberries and mushrooms after various types of freezing. (Sensoric test score: 1 = poor and 5 = excellent. FF = Fluid bed freezing; GF = Rapid blast freezing; LF = Freezing direct in liquid freon; LN = Freezing in liquid nitrogen; TF = Freezing in cartons in air blast tunnels). After Aaström and Löndahl (1969).

	FF/GF	LF	LN	TF
Strawberries				
Appearance	4.0	4.3	3.7	3.7
Taste	3.6	3.8	3.5	3.1
Texture	3.7	3.9	3.8	3.2
Drip (%)	16.2	18.7	21.0	26.7
Mushrooms				
Appearance	4.0	3.7	3.9	2.7
Taste	3.5	3.6	3.6	3.0
Texture	3.5	3.6	3.6	2.5
Drip (%)	15.3	16.6	15.1	15.0

Tables 14 and 117 show how very slow freezing of strawberries yields a product inferior to that obtained by quick freezing, e.g. plate freezing.

Table 15 shows an example of the effect of freezing rate on the drained weight of strawberries, i.e. a situation similar to but more pronounced than that often found for meat.

Table 15. Drained weight of ripe strawberries, "Surprise des Halles", frozen according to different methods. Lenartowicz, Plocharski, Zbroszczyk and Piotrowski (1979).

Freezing method	Drained weight %	Drip
Liquid nitrogen	83.2	16.8
Air blast	74.6	25.4
Plate	73.8	26.2

Problems of the effect of various freezing rates on the quality of frozen fruits and vegetables have been discussed in considerable detail by Ulrich (1981) and by Delaunay and Rosset (1981).

It must be repeated that where the product is cooked after freezing, the heat brings about texture changes very similar to those caused by freezing and often masks the effect of freezing on texture.

Bakery products. Most bakery products are very well preserved by freezing. Occasionally, the crust may tend to separate in the thawed product or firmness may be affected. Pence (1969) finds these phenomena practically independent of freezing rate. Yet some bakery products do seem to benefit from reasonably quick freezing.

Actual freezing rate

It is well to keep in mind that most processes, which are presented as quick freezing processes, are in fact only moderate freezing processes, because of slowness of heat transfer in freezing packages, larger cuts of meat, etc. Therefore, it seems wise to aim at a freezing rate from 0.2 to 0.5 cm/h, which will be possible in most normal freezing situations. Inefficient freezing, e.g. freezing packages bundled together in pallets, boxes in stacks without divider arrangements, etc., may result in such very low freezing rates that it may have an adverse effect on quality. This can very well be the case even where products are carelessly frozen in "quick freezing tunnels", blast freezers at -40°C, etc., and is mostly due to microbiological or autoenzymatic activity, which take place in those parts of the product which are farthest away from the surface, where the removal of heat takes place, cf. Fig. 17.

Any considerations regarding desirable freezing rates must be viewed in the light of what is actually achievable. In freezing beef quarters, blocks of fish, packages with beans, fluid bed freezing of fried potatoes, freezing shrimp in liquid nitrogen, etc., freezing rates are more or less determined by the type of equipment used. Since heat transmission rates are the main determinants, only limited increases in freezing rate can be achieved by lowering temperature. Table 16 shows some of the average freezing rates achieved in various conventional types of freezers, quoted by Olsson and Bengtsson (1972).

Table 16. Estimated freezing times and freezing rates in various types of freezers for different products. Olsson and Bengtsson (1972).

Product	Packaging	Weight	Freezing time hours	Average freezing rate cm/h
Air blast freezing:				
Potato chips	carton	400 g	4	0.55
Meat ball in gravy	carton	375 g	4	0.63
Raspberries,				
strawberries	barrels	40 kg	48	0.42
Meat, chopped	carton	30 kg	24	0.36
Plate freezer:				
Spinach	carton	425 g	2	1.1
Blueberries	carton	430 g	4.5	0.5
Steakburgers	carton	310 g	2	0.75
Cod filet	carton	300 g	1	1.30
Fluid bed:				
Peas	loose		7-8 min.	4.0
Belt freezer:				
Meat balls	loose	23 g	20 min.	4.5
Hamburgers	loose	60 g	20 min.	1.8

It is also necessary to consider that the freezing rate varies very much at various distances from the surface, cf. Figs 18 and 152.

The use of freezing in liquid freon etc., is likely to be discontinued because of environmental considerations.

Equipment efficiency

Attaining reasonably fast freezing rates has advantages quite different from product quality. Thus, it is often a great advantage if products to be frozen can be placed in a freezer on one working day, and the freezer unloaded early the next day. This may lead to better equipment utilization than slower freezing. Further, most continuous freezing arrangements, i.e. belt freezing, fluid bed freezing, automatic carton freezers, pellet freezers, etc., are based on a fairly

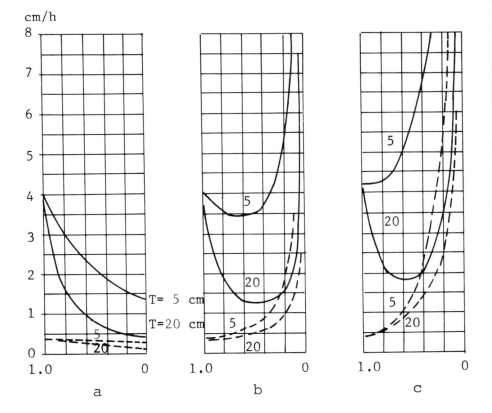

Fig. 18. Freezing rates at various distances from surface, in objects of different shapes. 0 = centre; 1.0 = surface. Distance from surface to centre either 5 cm or 20 cm. Heat transfer coefficient either 10 kcal/m²h. °C.---- or 100 kcal/m²h. °C.—. a) Plate shape, cooled from both sides; b) Cylinder shape; c) Ball shape.

rapid passage of the product through the freezer, i.e. from 30 minutes to 3 hours. Otherwise, the equipment becomes unwieldy, large and expensive.

Energy conservation and cost

Suggestions to increase freezing rates, as shown above, rarely serve a purpose in so far as improving product quality is concerned, although they mostly purport to do so. If followed, they generally lead to considerably added cost, a factor, to which little attention is often paid. First, such steps often call for much more expensive equipment and will necessitate a reduction of evaporator temperature. Figure 19 shows the characteristics of a two stage compressor. It is seen that lowering the evaporator temperature some 15°C will often nearly double energy requirements for the same heat removal, i.e. at a 20°C condensing temperature, and an evaporating temperature of -20°C, 29 kWh will remove about 94 000 kcal per hour, while at -35°C, 22 kWh will remove only 46 000 kcal. The additional need for refrigeration to compensate for additional heat transmission due to larger temperature differentials need also be considered.

In referring to Fig. 19, the author feels it appropriate to recommend that using warmer temperatures than normally employed be considered in some types of freezing equipment. Thus, consideration might be given to operating

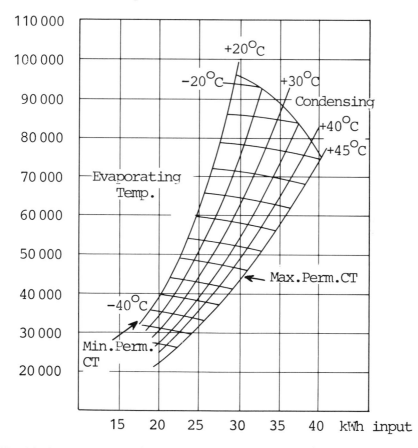

Fig. 19. Characteristics of a two-stage compressor as a function of condenser and evaporator temperature. CT = Condenser temperature. (Data from Sabroe A/S, Aarhus).

tunnel freezers and plate freezers at -30°C instead of the conventional -40°C. However, a calculation of the trade-off between equipment utilization and operation cost is indicated. The result will vary from case to case. Because of the constraints of product heat transmission characteristics, the freezing time may not be prolonged nearly as much as might be assumed.

In such calculations, a well considered choice of desired core temperature in the product when it leaves the freezer must also be agreed upon. This aspect is discussed below.

Woltersdorf (1982a) made available some preliminary data from experiments with freezing beef quarters at -18, -25, and -35°C, see Table 17. He found that the energy saved by freezing at -18 or -25°C was quite significant compared to energy use at -35°C. It probably could justify the loss of turnover rate in the freezers. He did state, however, that this was found for a freezer and compressor system designed to operate at -33°C, and concluded that redesign of the freezer installation might make freezing at somewhat warmer temperatures even more efficient.

One may also consider energy considerations in various freezing methods. Thus, cryogenic freezing is often advocated on the basis of improved flexibility and quality of product. The latter is commented upon above. Consideration should be given also to energy considerations. Here, Briley (1980) states that the energy requirement per 100 kJ removed is 0.404 kW for liquid nitrogen freezing, 0.106 kW for liquid carbon dioxide freezing versus 0.023 kW for mechanical freezing.

It is interesting to note that in Australia and New Zealand, freezer temperatures are seldom as low as believed necessary in North America and Europe, i.e. air blast freezing at -25°C is very common in the former countries.

In the view of the author, this is an aspect which deserves more study than has been devoted to it.

Core temperature

It is often required that a product not be removed from the freezer before the temperature in any part of the product has reached the stipulated storage temperature, often -18 or -20°C, see for instance UKAFFP (1978). Such demands will normally result in ineffective use of freezer installations, unnecessary complications for industry, and will have no meaning for product quality.

It is easy to see how this idea may have come about. When product storage temperature is specified, it seems obvious that this temperature should be reached by the product, i.e. in all parts of the product, before it is removed from the freezer installation proper.

Table 17. Energy use in kWh per beef quarter in freezing in a blast freezer at various air blast temperatures and various end core temperatures. Lean beef quarters weighing 95 kg were used. After Woltersdorf (1982b).

End core temperature, °C	Blast air temperature, °C		
	−18	−25	−35
−7	4.70	6.00	8.00
−10	4.75	6.05	8.50
−18	4.80	6.10	8.80

The explanation for lack of justification of this stipulation is illustrated in Fig. 20, showing temperatures in various parts of a satisfactorily frozen hind quarter of beef. It will be seen that the temperature in the outer layers is -31°C, while the core temperature is only +3°C. Yet, it serves little purpose to retain such a product in the freezer. By far the largest part of the block by weight has been reduced to a temperature below -20°C. In fact, if the product were left in an insulated box, temperatures would equalize and end up being -20°C throughout. Thus, all heat, which must be removed from the product, has been removed, and the product may be transferred without any adverse effect to a storage room at -20°C or even to a transport vehicle maintaining that temperature. This was discussed further by Fleming (1974), see also Fig. 21.

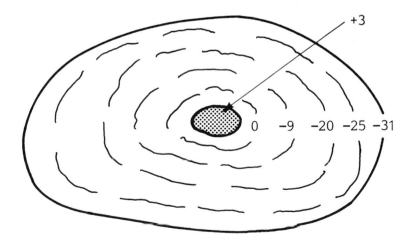

Fig. 20. Temperature conditions in °C in a pre-rigor frozen round of electrically stimulated beef at the time when heat has been removed sufficiently by freezing at -36°C, corresponding to an anticipated storage temperature of -20°C.

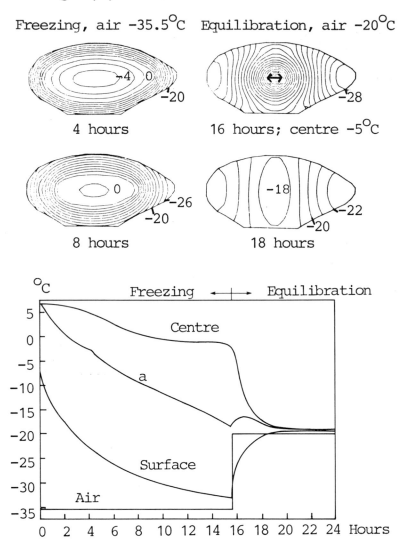

Fig. 21. Compiled temperatures in °C in a hind quarter of beef frozen at -35.5°C in an air blast of 4m /sec. "a" indicates temperature 4 cm under surface. Equilibration in still air at -20°C. After Fleming (1974).

Above indicates how freezing to a low end temperature may have an adverse effect on quality, cf. Winger and Pope (1981) and Fig. 12. Since a low core temperature normally is accompanied by quite low temperatures in the outer layers of the product as it leaves the freezer, this requirement may even have an adverse effect on quality, although it is often believed to be a means of achieving superior quality.

In Italy, a ministerial decree stipulates that what is termed deep frozen foods, must be frozen to a core temperature of -18°C in less than 4 hours. Since deep frozen foods probably refers to consumer-size packages, however, this stipulation hardly results in any adverse effect on the eating quality of most foods. As seen from Fig. 8 and Tables 6 and 7, it is also unlikely to result in any improvement.

Delaunay and Rosset (1980) indicate that in France a core temperature of -18°C must be reached as quickly as possible, i.e. in less than 1 hour for small pieces (steaks), 2 to 6 hours for medium sized pieces, e.g. roasts and poultry, and 12 to 24 hours for carcasses, and less than 24 hours for large packs, e.g. cartons with boned meat.

In a proposal for a directive on deep frozen foods, the Commission of the European Communities in 1982 suggested a very useful wording, namely that freezing should result in a "temperature which stabilizes at -18°C or colder in the thermic centre of the product". One may, of course, still wonder if it is reasonable to require that all products be frozen to -18°C or colder.

Quality Changes during Freezer Storage

Post freezing changes

As food science developed, what had been stated by Plank, Ehrenbaum and Reuter (1916) as mentioned above became increasingly clear, i.e. that most of the changes normally attributed to the freezing process were indeed unrelated to that process and, in fact, except for cases where texture is adversely affected by freezing, the frozen product is often practically indistinguishable from the fresh product when thawed immediately, cf. Table 18. However, after some months or even only weeks of storage, depending on product, process, packaging and storage temperature, changes will be noted, but it follows that these are due to changes during freezer storage rather than to the freezing process.

Table 18. Organoleptic tests of frozen and unfrozen chicken broilers. (Score: 0 = very poor; 10 = excellent). After Boegh-Soerensen (1977).

	Texture		Taste	
	Breast	Thigh	Breast	Thigh
3 h. after slaughter chilled	6.22	7.33	7.33	7.44
Frozen 3 h. after slaughter, thawed next day	6.94	7.22	7.08	7.33
Aged 2 days at 4°C	7.45	7.50	7.22	7.33
Aged 2 days at 4°C, frozen, thawed next day	7.56	7.36	7.19	7.39

Thus, as mentioned above, the freezing process was often considered the cause of thaw drip. However, in 1932, Reay (1933) carried out the experiments quoted in Table 19.

Table 19. Drip from minced haddock muscle, frozen at -20°C, subsequently stored at the temperatures indicated for 6 days, then thawed in water at 8°C. After Reay (1933).

Post freezing storage temp. °C	Average net drip, %
0	2.6
−1.2	14.3
−1.5	15.3
−3	16.3
−6	1.2
−12	1.8

He showed that drip after freezer storage at -12°C is quite small but that as little as one week of storage at -1.2 to -3°C causes excessive drip. The explanation generally offered is that the high ionic strength of the solution where most of the water has been frozen out causes rapid denaturation of proteins with lack of water binding as a consequence. This effect is not observed at colder freezer storage temperatures because of reduced reaction rates. In general, drip seems to be very much increased by warm freezer storage temperatures.

Lea (1930 and 1935) showed that lipids in fish oxidize quite rapidly during freezer storage. Similar findings were reported by Mullenax and Lopez (1975) as shown in Fig. 22, see also Table 20. Again, this illustrates how quality deterioration, in the early days associated with the freezing process, is due to processes that take place after freezing. The data also suggest that a very cold freezer storage temperature may not always be preferable. This conclusion should, however, be viewed in the light of products with neutral or reverse stability as indicated below and the packaging factor as discussed later.

Table 20. Time for formation of 0.5% free fatty acids in back fat of pig. After Andersen, Jul and Riemann (1966)

°C	Days
+25	1
+10	2.5
+5	6
0	9
−3	21
−20	120

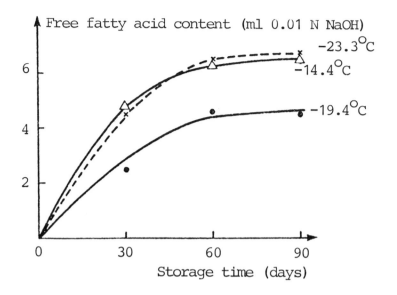

Fig. 22. Cumulative lipase activity in olive oil emulsion at three storage temperatures for an enzyme concentration of 0.10 mg per sample. Note that deterioration is less at -19.4°C than at -14.4°C or -23.3°C, i.e. a case of reverse stability. After Mullenax and Lopez (1975).

Findings such as those quoted above resulted in more and more research being undertaken to determine changes which take place during freezer storage, optimum freezer storage conditions, etc., rather than investigating the effect of the freezing process itself. In later years such problems have been the subject of most research on frozen foods. Yet, a vague idea of rapid freezing being essential has generally prevailed.

It is often discussed whether quality loss in foods, in this case frozen food, is a zero or first order function (Labuza, 1982). The curves in several shelf life determinations given below, e.g. in Figs 24, 27, and 28 suggest that several different types of functions exist, in fact it must even be accepted that different mechanisms may be at play and interact. Also, one type of degradation may be less offensive but occur at a more rapid rate and thus will not be the limiting factor in the beginning of the storage period but later turn out to be just that. The author finds he should warn against attempts to fit such quality loss curves into one or another mathematical pattern because this may often result in an assumption of an accuracy which is greater than what is justified from experimental data.

Time-temperature-tolerance (TTT)

Early years of experimentation into the effect of frozen storage on foods were characterized by intensive but somewhat unorganized data reporting. Results were reported like, "Peas of the ...... variety frozen for 20 minutes at -30°C will keep for 9 months at -20°C, and for 20 months at -26°C". Many results were obtained but they were somewhat difficult to compare, because various temperatures and various criteria for keeping quality were used. Data about initial product, process and packaging were often inadequate.

It was a great improvement when what is now designated the USDA Western Regional Research Center, Berkeley (earlier Albany, hence the expression the "Albany" tests), California, encouraged by the late H.C. ("Dutch") Diehl, undertook the so-called time-temperature-tolerance (TTT) experiments, cf. van Arsdel (1957). In the following, these experiments are referred to as the "Albany" series of tests. A great many different fruits, vegetables, and poultry products were tested at various freezer storage temperatures for various lengths of time. The quality was measured by various objective measurements such as ascorbic acid deterioration, the change of chlorophyll into pheophytin, colour, etc., and also - and primarily - by organoleptic testing. The latter was normally carried out by taste panels in triangle tests indicating when a change in quality was recorded. Results were normally reported in a uniform style in a semi-logarithmic diagram such as that reproduced in Fig. 23. The time was normally given on a logarithmic scale. As controls in subjective testing were used samples of the same product, stored at about -31°C or, at times, -40°C.

The series fell into two distinct groups, namely those where keeping times were determined by taste panel tests and those where objective criteria, e.g. 10% loss in ascorbic acid content, were used. In both cases, fairly straight lines in the semi-logarithmic diagrams were found for keeping time. This is remarkable since the unavoidable changes mentioned below in the controls should result in longer keeping times being recorded at colder temperatures where taste panel tests are used, while there should, of course, be no such effect where objective criteria are used since changes recorded in these are in relation to the freshly frozen product.

The "Albany" test series almost invariably resulted in straight line curves for the keeping quality of the products tested, i.e. a constant Q_{10} for the deteriorative processes during freezer storage, cf. Guadagni (1957). This was determined also by objective tests, e.g. chlorophyll conversion, ascorbic acid retention and objective colour measurements, cf. Olson and Dietrich (1969). However, this seems to be characteristic for the products and temperature range included in the experimental design. As discussed below, in other cases, and especially for other products, vastly different shapes of the keeping time curve were found, *viz.* Figs 35 and 36.

It is interesting that in most cases Q_{10} is of the order of 10 while it for many simple chemical reactions is in the order of 2-3, probably somewhat related to the more complex processes which take place in the solidly frozen tissue.

While it is often assumed that the "Albany" series of tests always resulted in straight lines in a semi-logarithmic diagram, Fig. 24 shows that this almost by the very nature of the experiments cannot always have been so if overall quality is considered. Most tests were carried out by recording quality loss by a taste panel. A panel will normally be influenced by its overall impression of the samples.

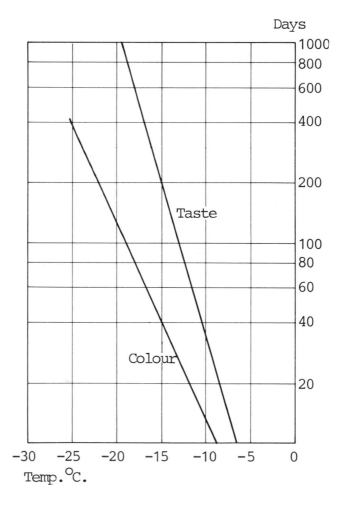

Fig. 23. Stability times for cauliflower determined in the Albany test series. After Dietrich *et al.* (1962.)

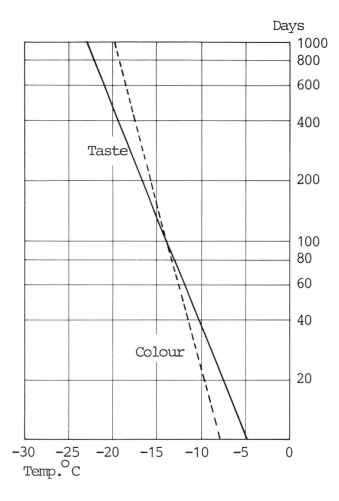

Fig. 24. Stability time for peaches. It is noted that if overall quality is considered, the existence of two different limiting factors may result in a TT-curve which is not a straight line. After Guadagni, Nimmo and Jansen (1957a).

Figure 24 shows that for peaches at warm temperatures colour changes are the limiting factor, at colder temperatures it is taste. It is seen from this that if overall acceptability is considered, a broken curve is to be expected, even if a constant Q-value applies to each change reaction in the product. As will be indicated below, curved lines are the rule rather than the exception for the various keeping quality diagrams, normally referred to as TTT-diagrams, cf. Fig. 25 for pork, and Fig. 26 for meat balls, both determined by Lindeloev (1978).

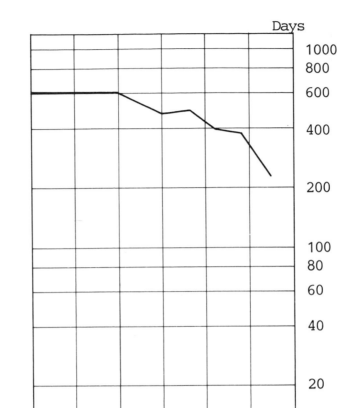

Fig. 25. Shelf life (acceptability times) for vacuum packaged, untreated, sliced pork, determined by Lindeloev (1978).

Triangle testing technique

As mentioned above, a methodology used at the "Albany" investigations was that of comparing samples at the test-temperatures with controls at -30 or -40°C. The end-point of the keeping time was recorded as the time elapsed when a taste panel in a triangle test registered a stastistically significant change between the controls and the test-sample. As a standard was taken that the change was detected with a probability of 95%. It is quite evident that this system worked well for the products and temperature ranges considered at

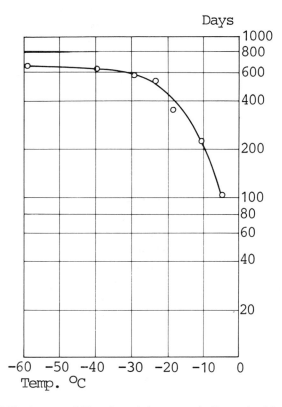

Fig. 26. Shelf life (acceptability times) for meat balls packed in polyethylene pouches, without vacuum, determined by Lindeloev (1978).

Albany, i.e. fruits, vegetables, and poultry at temperatures down to -20°C; also, as mentioned elsewhere, it was carefully checked and also compared with objective tests, cf. Guadagni (1957). One would, however, have considerable difficulties with this method for products having a keeping quality character-istic like the ones shown in Figs 25 and 26. Here, the controls will undergo considerable change during the storage process. This means that the test samples are not being compared with an unaffected product which in a way should be the fresh product or at least one just frozen and thawed; obviously, questions of seasonability, differences in varieties, etc., prevent the latter from being a usable avenue.

Thus, the results of the Albany tests series in practically all cases resulted in a straight line in a semi-logarithmic diagram, i.e. a constant Q_{10}. Later tests have shown somewhat curved lines in most such diagrams for both stability and acceptability. One may wonder if some of the relatively long shelf lives, which were reported in the Albany test at low temperatures, could not to a certain extent be an artefact, i.e. the result of the fact that the controls also change. Obviously, if the controls deteriorate during cold storage although at a lower rate than at warmer temperatures, this is likely to result in less difference between test samples and controls, i.e. longer stability times. This effect will, of course, be particularly pronounced the nearer the storage temperature of the test samples approaches the temperature of the control.

Time of comparison

Another question has been raised as to the methodology used in some of these experiments. Test samples are normally compared at various times with a control kept at -30 to -40°C. However, the underlying assumption that products do not change at all at -30 to -40°C is, of course, only approximately correct. Then, a test product stored for a long period obviously gets compared to a different sample than that used as control for a sample stored only for a brief period. This objection could be overcome if all samples were kept for the same length of storage time. For instance, if various storage periods at -12 and -15°C are to be tested, one could store the test samples at those temperatures at various specified times and then leave them for the remainder of the period at -40°C.

In accordance with this, McBride and Richardson (1979) pointed to the usefulness of testing all samples in an experiment against uniform controls using this technique. They transferred all samples tested at higher temperatures to the storage temperature for the controls, in this case -30°C, until all samples could be tested at the same time.

Guadagni (1957b), in testing the validity of the experimental design, made experiments in this regard also by storing control and all test samples for the same period. For the test samples, make-up freezer storage was at -29.9°C, i.e. the same as for the controls. For the product range included in the "Albany" experiments, no need was found for such extra precautions.

The latter technique could have the disadvantage that many samples are to be tested at the same time, which technically can be inconvenient. Further, it assumes that the effect of a storage period is independent of the prior history of the product, i.e. it is assumed that a product, which has been stored for some time at -12°C, will change at the same rate at -40°C as one which has been stored at that temperature throughout. As is indicated below, it seems as if this assumption is generally justified.

In connection with the "Albany" series of tests, this matter was, of course, given careful consideration. Thus, Dietrich *et al.* (1959) did determine, also by objective tests, e.g. chlorophyll conversion, ascorbic acid retention and objective colour measurements, that changes during storage at -28.9°C were negligible.

The conclusion must be that the methodology described above was suitable for the products for which it was used but must be closely examined for other products. Used uncritically it can give misleading results.

Stability and acceptability time

In interpreting the results from such experiments, one should consider also another factor. When an experienced taste panel carries out a triangle test, only very minor differences are required for the panel to record a change with great probability. Thus, this is a very exacting test. It means that the keeping quality times recorded according to this method cannot be those that would apply under commercial conditions. The author has referred to the keeping times found in such experiments as the time to *just noticeable difference,* JND, or *first noticeable difference,* FND, or *stability time.* It has also at times been referred to as time to *first perceptible difference,* FPD. The distinction between this and the keeping time which applies in practice has, for instance within the framework of the International Institute of Refrigeration, IIR, given rise to discussions. Around 1960, it was suggested that in regulations pertaining to the keeping quality of a frozen product, one should use the concept of time to first noticable difference. In IIR publications, this is referred to as a product's *high quality life,* HQL. It was, however, maintained by the author that recommending product storage time in this way would be unrealistic. One would arrive at recommended maximum keeping times so short that a considerable share of frozen products as they are sold commercially today would be considerably beyond the prescribed maximum keeping time. This would conflict with the fact that consumers in general accept the frozen products as they are found in the trade today and have few quality complaints. Obviously, the trade and consumers are not nearly as sensitive to small changes as is a taste panel, especially where the latter use triangle tests. This led to the introduction of what has generally been referred to as the concept of practical storage life, PSL. The author uses the term, *acceptability time.* It seemed obvious that for practical applications, one would have to operate with such a keeping quality time concept, namely the maximum time where a product would remain fully acceptable to a discerning consumer. This concept, although not necessarily the term, has been generally accepted. One might simply refer to this as the shelf life of a product. The difference between stability and acceptability times was discussed later by Guadagni (1969) and by Olson and Dietrich (1969).

In the view of the author, the concept of acceptability needs to be studied further, taking into consideration not just what is recorded by taste panels but also what is accepted by and acceptable to consumers. Appearance is one factor which needs to be taken more into consideration in experiments.

Table 21. Stability times, i.e. time in days for a decrease in flavour scoring of approx. 1 point approximately indicating first detectable loss in flavour. (This varied from a decrease in score of 0.8 to 1.1 points for the different products.) Maximum storage time tested was 160 days at -5°C and 320 days at -10°C and -20°C, except for products 10, 11, and 12: 80 days at -5°C, 160 days at -10°C, and 320 days at -20°C. Calf liver I was a very fresh product, calf liver II had gone through a moderate period of cooler storage. (Scores: -5 = very poor; +5 = excellent). After Dalhoff and Jul (1965).

Product	Temperature		
	-5°	-10°	-20°
1. Steakburgers	16	210	250
2. Rumpsteaks	20	230	>320
3. Calf liver II	24	52	170
4. Hamburgers	17	64	135
5. Ground beef	51	120	>320
6. Pork chops	27	105	180
7. Ground pork	41	79	230
8. Pork liver	116	100	170
9. Calf liver I	60	165	>320
10. Pork cooked with rice and curry sauce:			
The meat	>80	>160	>320
The sauce	>80	>160	>320
11. Pork liver cooked with cream sauce:			
The meat	56	75	230
The sauce	65	130	(420)
12. Hamburger cooked with onions and sauce:			
The meat	>70	>160	>320
The sauce	>80	>160	>320
13. Pork sausage	22	110	240

Thus, the appearance of frozen ground beef in the frozen state is seriously affected by poor packaging. The effect is almost completely eliminated in cooking, e.g. the preparation of samples for presentation to taste testing sessions. Thus, any meaningful evaluation of the effect of packaging must include an evaluation of appearance in the frozen state since this may be very decisive for the consumer's decision as to whether to buy the product or not.

The author is aware of the fact that no clear definition exists of acceptability. One reason is that different markets and different groups of consumers may have different expectations. Also, the satisfaction of the consumer is the only true criterion to use, but it is nearly impossible to measure. However, to replace this by some other definition, e.g. objective measurements, would be to disguise the issue. Conversely, the use of both laboratory taste panel and, at times, some objective tests will in the hand of experienced persons with close contact to trade and consumers yield results that are reasonably satisfactory - and the only measure available.

Table 22. Acceptability time, i.e. time in days for a decrease in flavour scoring to a score of -2 for the same products as in Table 21. A figure in brackets means that the decrease to a score of -2 was not actually found during the testing but was estimated by linear extrapolation on a diagram plotting flavour scores against days of storage. Double bracket means that not even a decrease to a score of -1 has been established. One must assume that the actual keeping times are longer than those stated in the brackets and may be much longer than the figures in double brackets indicate. "No decrease" means no decrease found after 80 days at -5, after 160 days at -10, and after 320 days at -20°C. For calf liver I and II, see Tables 21 and 52. (Scores: -5 = very poor; +5 = excellent). After Dalhoff and Jul (1965).

Product	Temperature		
	–5°	–10°	–20°
1. Steakburgers	57	((650))	((750))
2. Rumpsteak	72	((690))	no decrease
3. Calf liver II	60	135	(340)
4. Hamburgers	56	(242)	(391)
5. Ground beef	(163)	((396))	no decrease
6. Pork chops	86	(347)	((610))
7. Ground pork	(106)	253	((645))
8. Pork liver	(197)	220	(418)
9. Calf liver I	96	297	no decrease
10. Pork cooked with rice and curry sauce:			
The meat	no decrease	no decrease	no decrease
The sauce	no decrease	no decrease	no decrease
11. Pork liver cooked with cream sauce:			
The meat	((185))	(180)	(460)
The sauce	((253))	((442))	((1300))
12. Pork liver cooked with onions and sauce:			
The meat	((252))	no decrease	no decrease
The sauce	no decrease	no decrease	no decrease
13. Pork sausage	57	275	(480)

Moleeratanond *et al.* (1981) found only very limited correlation between a panel type organoleptic evaluation and scoring by a consumer type panel.

Using as a target acceptability among Danish consumers, Dalhoff and the author (1965) determined stability and acceptability time for a considerable number of meat products as seen in Tables 21 and 22.

Lindeloev (1978) similarly determined these data for many other meat products, and Boegh-Soerensen and Hoejmark Jensen (1981) determined those shown in Figs 37 and 38. However, as mentioned later, in these cases an end score of -1 was used as criteria for acceptability.

Boegh-Soerensen (1975) determined similarly such data for chicken broilers and chicken parts as seen in Figs 29 and 30.

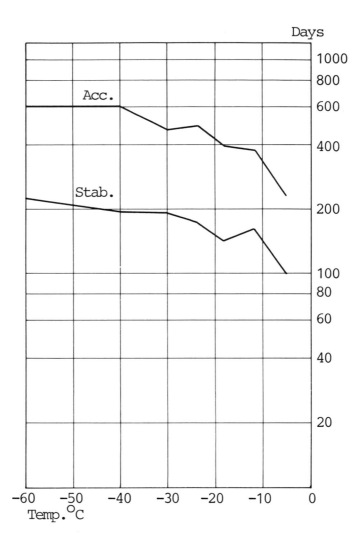

Fig. 27. Stability and acceptability times for sliced, vacuum packaged pork belly meat. After Lindeloev (1978).

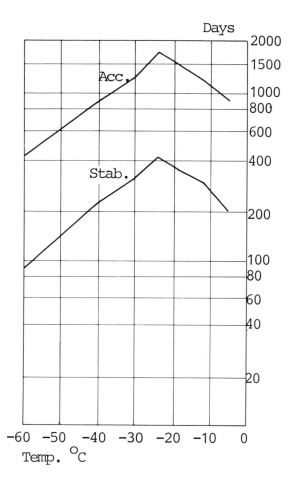

Fig. 28. Keeping times for vacuum packaged wiener sausages. After Lindeloev (1978).

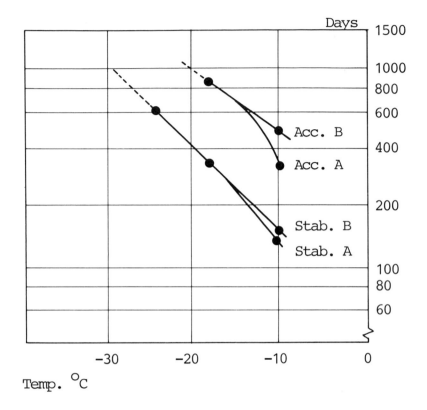

Fig. 29. Stability and acceptability of broiler chickens, A. in 0.05 mm PE-bags, B. in 0.10 mm bag of nylon/PE-laminate or cry-o-vac. After Boegh-Soerensen (1975).

Acceptability factor

In many cases it may be difficult to determine both stability time and acceptability time, simply because many samples, many tests and a long time will be required. Actually, while mostly data on acceptability times are needed, it is often easier and more convenient to determine stability times. The illustrations just referred to give for several products the relation between these two different curves. Out of such studies has come one interesting feature. It does appear as if the ratio between acceptability time and stability time is more or less constant over the usual range of temperature for any one product type. Depending on the degree to which this is correct, one might operate with a concept for which the author has used the term *acceptability factor*, i.e. one may determine for each type of product this ratio and in the

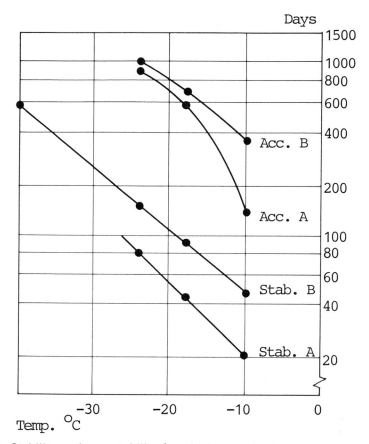

Fig. 30. Stability and acceptability for chicken parts; A. on plastic tray and wrapped in PE-film, B. in vacuum-bags. After Boegh-Soerensen (1975).

interpretation of experimental data assume that one can arrive at the acceptability time by multiplying stability time with the acceptability factor. For some products the author, cf. Andersen, Jul and Riemann (1966), has concluded that the acceptability factor could be as indicated in Table 23.

As mentioned later, this factor may be useful in accelerated storage life tests. However, some caution has to be used in interpreting this concept. Figures 29 and 30 show cases where the factor is not constant over the usual temperature range for products packed in permeable films.

Nevertheless, the factors quoted in Table 23 modified by experience in each case may be found useful in making estimates of acceptability times when stability times are known. It will be recalled that determining acceptability times even just down to as cold as -20°C may be very time consuming, i.e. take about three years for a frozen poultry product, cf. Fig. 29.

Table 23. Acceptability factors, i.e. ratio between acceptability and stability time for some product groups.

Beef	3-5
Pig meat	2.5-3.5
Cured, vacuum packaged	3.5-5.0
Cured, packaged in permeable material	2.0-3.0
Chicken, whole broilers	4.5
Chicken parts	6-9
Lean fish	2
Vegetables, except peas	4-15
Peas	6-15

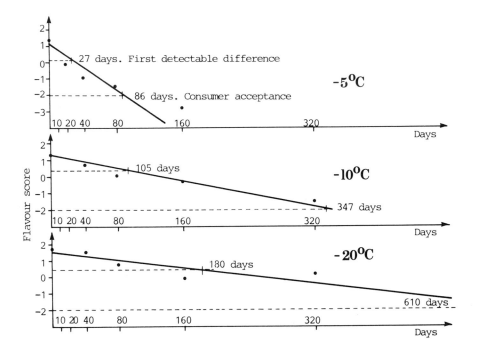

Fig. 31. Determining "first detectable difference", i.e. stability time, and acceptability time for pork chops (score: -5 dislike extremely, +5 like extremely). (Dalhoff and Jul, 1965).

Scoring method for keeping quality determination

Above are discussed some of the problems inherent in the method used for determination of the keeping quality in the USDA Western Regional

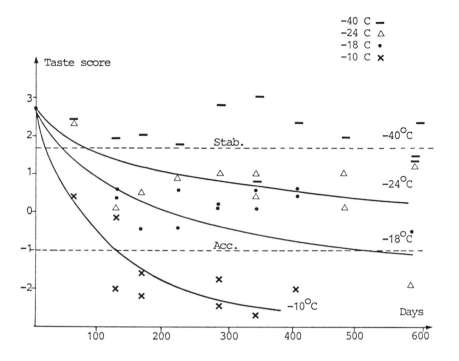

Fig. 32. Taste scores for broilers including samples kept at -40°C. (Score: -5 extremely poor, +5 extremely good. (Boegh-Soerensen, 1975).

Research Center. To what is discussed above, one might add yet another comment. This method compares one frozen product with another, i.e. the frozen controls. One may then have certain problems in really deciding the relation between stability (HQL) and any concept of shelf life. For instance some products are really not acceptable once they have been frozen, e.g. bananas, tomatoes, etc. The concept of comparing with a control stored at a cold freezer temperature would lead to confusing results if one compared one unacceptable product with another and noted the time when a difference between the two was detected. Data from such experiments would be meaningless, at least if interpreted in the usual way.

In the late 1950's, a different method of determining stability and acceptability times was introduced at the Danish Meat Research Institute, Roskilde. It was first described by the author at the IIR Congress in Munich in 1963 (Dalhoff and Jul, 1965). The samples to be tested were presented to a taste panel specializing in testing the type of products under examination. The panel was instructed to give scores using the same criteria they used for unfrozen products; in fact no effort was made to inform them that frozen products were being tested, although this fact may often not have remained

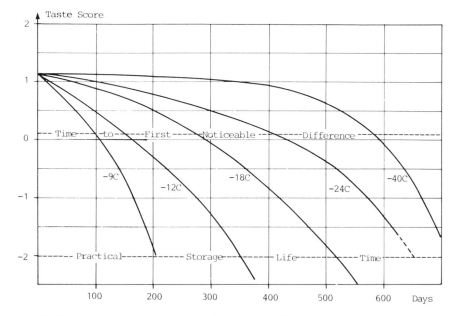

Fig. 33. Results of taste tests for pork chops at different storage temperatures. Taste score: -5 dislike extremely, +5 like extremely. Packaging: wrapped in polyethylene in a freezer carton. (Boegh-Soerensen, 1967).

undisclosed for long. However, the panel was experienced and recorded its opinion about the products using the criteria it normally used for other products of the same type, either frozen, canned or fresh, i.e. in a -5 to +5 scale, where -5 was dislike extremely and +5 was like extremely. Results were such as indicated in Figs 31 - 33.

It was found that a reduction in average scores of 0.8 to 1.1 points in the 11 point hedonic scale used corresponded quite well to the stability time as determined by the "Albany" method. This can, of course, mainly be ascertained for such products where the "Albany" method can be used. Thus, it was done in the "Albany" test series for peas and poultry, which seem to have good quality retention at -40°C. Figure 32 suggests that controls maintained at -40°C will constitute a reasonable means of evaluating change for chicken.

However, the author and co-workers used another criterion, i.e. in the first years a score of -2, which was considered the limit for acceptability. Results obtained at the Danish Meat Research Institute, Roskilde, and at the Danish Meat Products Laboratory, Copenhagen, have since been recorded according to this system, which seemingly has given satisfactory results. These laboratories concern themselves mainly with meat products, where, as indicated above, the "Albany" method will often not be meaningful, e.g. in cases of

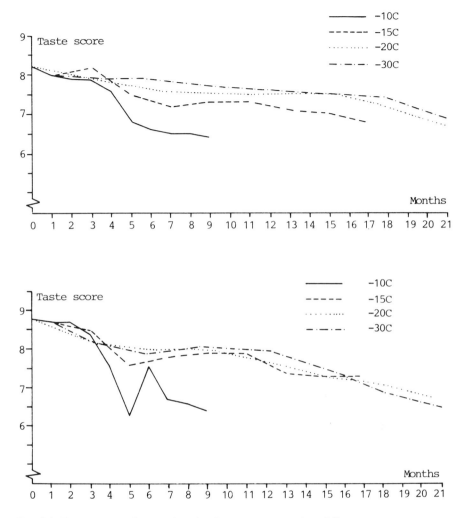

Fig. 34. Taste scores for chicken broiler meat, stored at different temperatures. Top: breast meat; Bottom: thigh meat. (Scale: 1 objectionable, 9 excellent). After Ristić (1980a).

neutral or reverse stability as described below, see also the rapid change of a sample kept at -40°C in Fig. 28. After some years, the limit for acceptability was changed to a score of -1.

Dr. Peter Howgate, Torry Research Station, Scotland, has called attention to the fact that any scoring system, whether it be intensity systems, i.e. scoring for intensity of flavour, degree of toughness, etc., or a hedonic scales, e.g. like extremely, dislike extremely, etc., may be interpreted in different ways.

Generally, one uses an average of the scores given by a number of panel members. However, it may be incorrect to assume that the shelf life is indicated as the time when the average score given by the taste panel reaches the point of no longer being fully acceptable. It may be argued that at this point about 50% of the panel is likely to have found the samples no longer acceptable. Thus, one should rather use that point where for instance 5% or possibly 25% judge the sample no longer acceptable, see time-temperature surveys. While this in principle is correct, it leads to a degree of sophistication of the interpretation of taste panel results which is difficult to implement. At any rate, any definition of limit of acceptability is so arbitrary that one may well be as good as another. The system as used by the author and his colleagues has given quite satisfactory results, cf. the Conclusion of calculations section below.

Ristić has, for chicken, developed similar curves to those in Figs 31-33, cf. Fig. 34 and, in (Ristić 1980b), somewhat similar curves for turkey meat. Kramer, Bender and Sirivichaya (1980) report some cases of an S-type of curve in similar taste tests of frozen foods. However, such shapes of the curves would not affect their utility, at least not for determining acceptability times.

Where the scoring system for determining keeping quality results in straight lines or curves as in Figs 31-33, the acceptability factor seems to be fairly constant, i.e. the same ratio between acceptability time and stability time applies over the whole temperature range tested. This is not always the case as seen in Figs 29 and 30. From these it is clear that the acceptability factor may be higher at cold temperatures than at warm temperatures for products packed in oxygen-permeable film.

Reverse stability

In the mid-fifties, Danish bacon factories occasionally had to resort to freezing Wiltshire bacon, that is cured pork sides, because of market conditions in the UK. This led to some peculiar results. Complaints were received about those sides which were frozen in very modern cold storage warehouses, while products frozen and stored in very old plants, generally considered less efficient, caused no problems. The problem was investigated by the author who concluded that what was considered the old-fashioned freezer installations actually never brought the temperature down to the freezing point, i.e. about -4°C. Since it was known that freezing may accelerate oxidation, it was concluded that the problem was accelerated rancidity which was due to storage in the frozen condition *per se*.

This conclusion was based on the observation that the "modern" plants

Table 24. Acceptability times in months for vacuum-packaged cured pork products. Accceptability time measures the time until a score of -2 is recorded on the same -5 to 5 points scale as used in Fig. 31. After Boegh-Soerensen (1968).

	-6°C	-19°	-23°
Danish cured pork roll	>9	>9	8-9
Smoked sliced streaky bacon	>6	2-6	6
Unsmoked sliced back bacon	8	2-9	8

Table 25. The stability of sliced bacon in oxygen permeable film and according to rancidity (+ = rancid, (+) == trace of rancidity, - = no rancidity). (Nilsson and Hällsaas, 1973).

Days	2°	-5°	-12°	-18°	-25°
4	–	–	–	–	–
8	–	–	–	–	+
11	–	–	(+)	(+)	+
16	–	(+)	+	+	+
21	–	(+)	+	+	+

froze the Wiltshire sides hung individually in air blast freezers at -40°C, while the "old" plants placed the sides, packed four and four in burlap wrappers, on top of each other, and in freezer storage room of -10 to -12°C.

Some ten years later, Boegh-Soerensen (1968) reported some experiments with frozen vacuum-packaged cured sliced pork products, cf. Table 24. Some few samples gave very short acceptability times, presumably because of leakers. For the rest, the acceptability time seemed almost independent of whether freezer storage took place at -6, -19 or -23°C, smoked streaky bacon even seemed to keep better at -6 than at -23°C. In the experience of the author, this seems to be the first time where definite evidence was presented of the fact that lowering freezer storage temperature may lead to reduced keeping time.

Five years later, Nilsson and Hällsaas (1973) of the Swedish Meat Research Institute, produced the data given in Table 25.

These data very clearly indicate that frozen bacon in oxygen-permeable packaging, here oxygen permeable film, keep longer stored at -5°C than at -18°C, and that the keeping quality is longer for a non-frozen product. Lindeloev (1978) points out that it is well-known that several pharmaceuticals keep better above than below freezing temperatures. However, the least expected result, which was obtained by Lindeloev, was that at very low temperatures, -25°C or colder, the keeping quality of bacon may be even poorer than at -5°C. This so-called reverse stability was subsequently

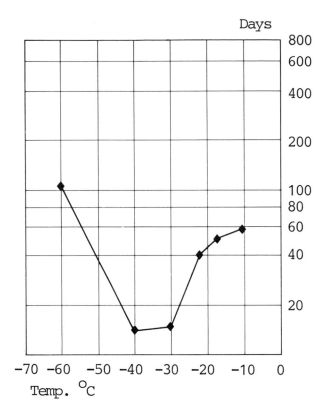

Fig. 35. Stability time in taste tests for sliced smoked bacon packed in oxygen-permeable film (polyethylene), reported by Lindeloev (1978). The product shows extreme reverse stability down to -30°C.

extensively investigated by Lindeloev. It seemed important to consider such cases further, because the conventional concept of storing frozen foods has hitherto been based on the tacit assumption that a colder temperature always would result in an improved keeping quality, which obviously is not true for all products. Figures 35 and 36 show some stability curves obtained by Lindeloev for cured pork.

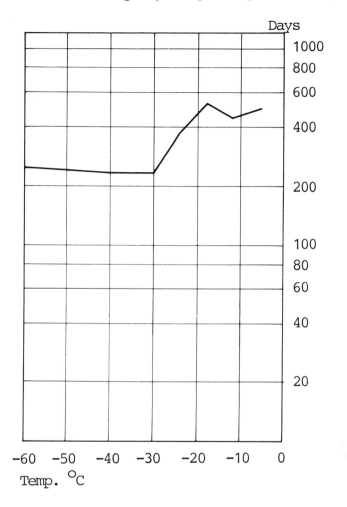

Fig. 36. Acceptability times for sliced vacuum-packaged bacon by taste tests, reported by Lindeloev (1978). The product shows moderate reverse stability.

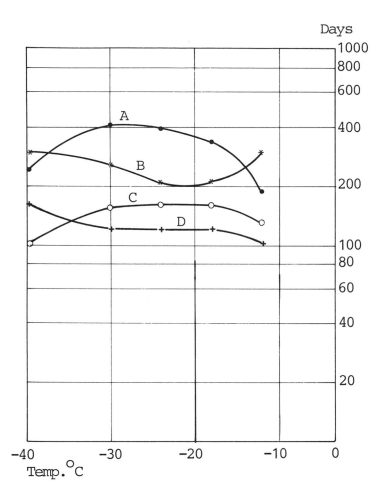

Fig. 37. Stability time for vacuum packaged sliced smoked bacon. Two salt/water ratios, either 7 g (A and D) or 3 g (B and C) NaCl per 100 ml H_2O and two plastic materials with oxygen permeabilities of 10 (A and B) and 1200 (C and D) ml/m^2 x 24h x atm. respectively were used. (Boegh-Soerensen and Hoejmark Jensen, 1981).

Other cases are shown in Fig. 37 for sliced bacon and in Fig. 38 for a mixture of ground beef and ground cured pork. All are time-temperature characteristics very different from those conventionally assumed to apply for frozen foods.

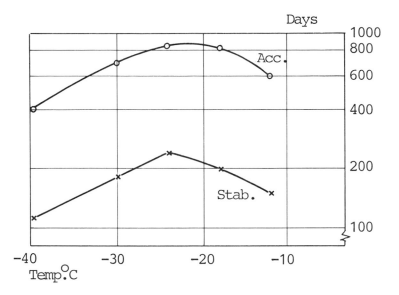

Fig. 38. Keeping times for breaded bacon burgers packaged in coated paperboard cartons. (Boegh-Soerensen and Hoejmark Jensen, 1981). The product shows moderately neutral stability.

Neutral stability

Figure 39, also determined by Lindeloev (1978), as well as Figs. 37 and 38 show that, at least at temperatures around -18 to -30°C, the shelf life of some products is practically independent of temperature, a phenomenon referred to as neutral stability.

In one case, Ristic (1982) reported that no difference was observed in storage life of chicken broiler parts stored at -15°C and -21°C. When retention of B-vitamins in frozen meat was measured, no difference between storage at -12 and -26°C was found in several cases, i.e. this attribute seems to display neutral stability in this temperature range, cf. Tables 44 and 50. Similarly, Klose *et al.* (1955) found no beneficial effect of storing well wrapped turkeys at -35°C compared to -23°C.

For thiamin and ascorbic acid content in peas and broccoli, Kramer (1979) finds only marginally higher losses in storage at -10°C compared to storage at -30°C, i.e. another case of apparent neutral stability. Molecratanond *et al.* (1981) also finds little effect of temperature on the retention of thiamin, riboflavin and niacin, see also Table 52. While retention of some B-vitamins seems

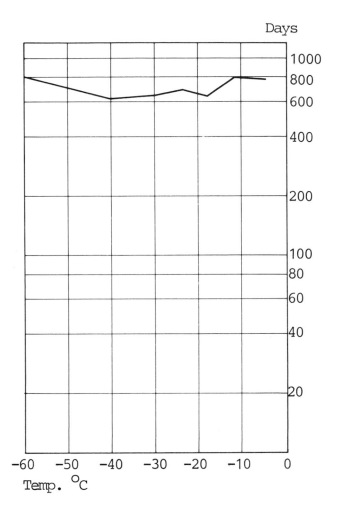

Fig. 39. Acceptability times for smoked, vacuum-packaged bacon, determined by taste tests. After Lindeloev (1978). The product shows neutral stability over the whole temperature range tested.

largely unaffected by storage temperature, many findings, e.g. many for the "Albany" test series, suggest that retention of ascorbic acid is significantly improved by colder temperatures.

Determining shelf life

It is, of course, very important that a producer of frozen foods determine the keeping quality of the products in order that approximate date marking and decisions as regards distribution options may be made. Sanderson-Walker (1979a) indicates how this is done by Birds Eye Foods Ltd., UK. He indicates how a relatively short term predictive test for storage life is carried out by testing at -12°C for 12 weeks. The rationale behind such accelerated tests, which were suggested by the author, see Jul (1960), is discussed below.

Accelerated tests

Curves such as those given in Figs 23, 24, 29 and 30 suggest that one could use accelerated tests for testing the keeping quality time of frozen products, and with a considerable degree of accuracy predict the keeping quality of a product at colder temperatures. This is very important because changes in packaging, processing, etc., as discussed in detail below, may have a large effect on the keeping quality of a frozen product. However, if the magnitude of the effect of a modification in product process or packaging cannot be known until after a test of one to three years, another change in product or packaging may by then be contemplated, and one would never have any clear idea of the actual effect of such changes on the keeping quality of a product until it was too late. Since the shape of time-temperature-tolerance curves has been generally determined for a wide range of products and may be determined with greater accuracy for any specific set of conditions, e.g. one manufacturer's usual products, one may carry out accelerated tests at temperatures as warm as -10°C or even -8°C, and may thus, often in a few weeks or months, be able to predict keeping quality at the anticipated storage temperatures, e.g. -20°C.

When looking at some of the TTT-diagrams, one might be tempted to use even warmer test temperatures, and this may be possible for some fruits and blanched vegetable products. For other products, e.g. meat or fish, this is hardly practical because microbiological activity may not be halted until -12°C or even colder temperatures are reached. Actually, mould growth has been recorded down to -17°C. On the other hand, Lowry and Gill (1982) found no evidence of microbiological growth on meat products at -8°C. Thus, this, or, as Winger (1982) suggests, even -5°C may be a safe test temperature for accelerated tests for stability of frozen meat. However, spoilage may set in before the acceptability time is reached. Therefore, in determining acceptability times, the author generally recommends -8°C as the warmest temperature.

The above-mentioned acceptability factor may be a useful concept for accelerated tests. First, it may be easier to determine the point of just noticeable difference, i.e. stability, than the limit of acceptability in experi-

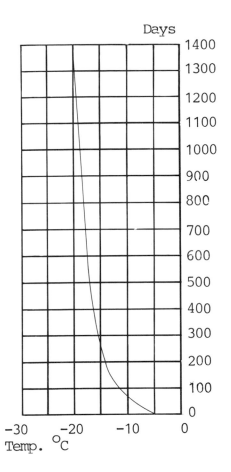

Fig. 40. The stability time of cauliflower, according to taste, using a geometric time scale.

mental procedures. Second, this point may be determined at relatively warm temperatures, even -5°C for acid or blanched foods, without interference of microbiological activity. These two factors may mean that keeping times can be predicted after reasonably short experimental periods. Thus, such as shown in Fig. 30 for chicken parts, the superiority of vacuum-packaging might be determined after about 10 days at 10°C when the polyethylene packaged product will show first noticeable signs of change while the vacuum-packed one will not. After 50 days, when the well packaged product begins to show signs of change, one will know that that product is likely to store about 3 times longer than the other. One must be aware, of course, of such phenomena as shown in Fig. 30 where an improvement at warm temperatures is not necessarily parallelled by a similar difference at the most likely storage temperature, or where a slight

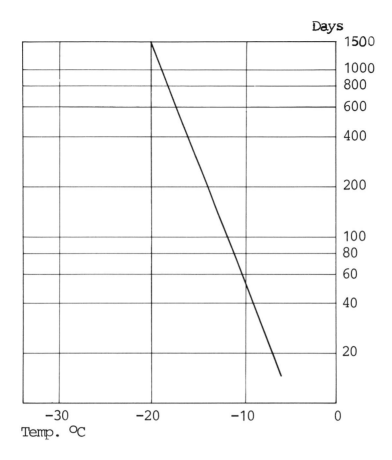

Fig. 41. The stability time of cauliflower, according to taste, using a logarithmic time scale, cf. Fig. 23.

product modification, e.g. the addition of salt to meat products, cf. Fig. 49 compared to Table 60, changes the keeping characteristics completely.

The logarithmic scale

The Western Regional Research Center's style of diagrams with a logarithmic scale for time was probably inspired by the fact that using a geometric scale for time in tests with fruits, vegetables or poultry would result in a somewhat cumbersome diagram such as that given in Fig. 40, which applies to the same product as Fig. 41.

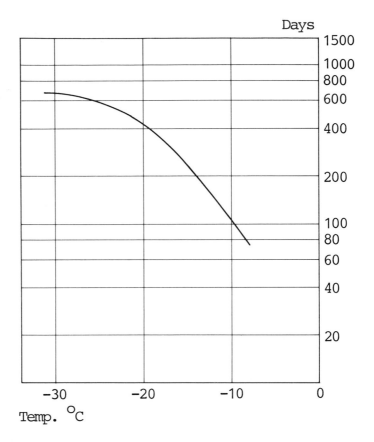

Fig. 42. The stability time of pork sausage, according to taste, using a logarithmic time scale.

As explained in relation to Fig. 24, one will notice that for most products a linear curve is not obtained. When overall quality is measured one will almost invariably find bent shapes of the curve.

Especially where meat or fish is concerned, the curves in a semi-logarithmic diagram look more like Fig. 42. For these products, geometric scales may appear more appropriate. In that type of diagram, the same product would be characterized by the curve in Fig. 43.

Thus, Spiess (1980) in an unpublished report prepared for Codex consider-ations, giving keeping times for a great many products, assembled from literature data, used a geometric scale, cf. Figs 69 - 74.

The author personally prefers the use of a semi-logarithmic scale. Even if only few products exhibit a linear shelf life in such a diagram, the diagram is

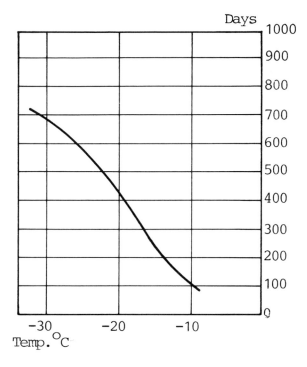

Fig. 43. The stability time of pork sausage, according to taste, using a geometric time scale, cf. Fig. 42.

convenient for products with long keeping times at low temperatures and is even quite practical for products with neutral or reverse stability, cf. Figs 35 -39. Thus, the choice of a logarithmic type of diagram is convenient for products with very long keeping times, and it is not inconvenient for any other type of product.

This choice is not due to any assumption that the quality loss in frozen foods has an Arrhenius-type time-temperature relationship. In fact, the author doubts if a truly exponential time-temperature shelf life curve exists for frozen foods, although Strachan (1983) still assumes such a shelf-life function to be normal. Attempts to fit TTT-curves to straight lines or any other simplified function are to be warned against, especially since this may lead to very convenient and convincing mathematical calculations which because of the variability of the biological material will be crude approximations only, a fact easily disguised by the precision of the mathematical treatment. Further, any calculation as regards shelf life loss is so easily carried out on the basis of experimentally determined curves that such mathematical treatment is of little advantage.

Standardizing taste panel evaluations

The "Albany" system, i.e. triangle tests for first noticeable difference, has one feature which is very attractive, namely that it permits quite exact comparisons between experiments in different countries and at different times, because a first noticeable difference is likely to be detected by practically any trained taste panel at the same point regardless of local preferences. A scoring method, e.g. using the scale indicated in Fig. 31, may be quite population specific. For instance, the author was at one time involved in comparing by taste-tests iced cod in Denmark with the same type of fish and test in the United Kingdom. This was at a time when most cod in the UK was trawler caught and had been up to 10 days in ice before it reached port, while the Danish consumer would not consider a fish fresh unless it was brought home alive or at least was still alive when purchased by the consumer. It is quite clear that quality concepts would be different between two populations with these experiences, and it would be unrealistic to try to harmonize taste panels in these two locations because they would have to disassociate themselves completely from the acceptability standards with which they normally worked.

The author personally maintains that this is not necessarily a valid objection to scoring tests, however. The ultimate purpose of most experiments with frozen products is to find out how fully acceptable products are best brought to the plates of the consumer. Thus, acceptability has to be judged against the expectations of the consumer. The fact that these are different in different situations, e.g. different countries, is not something which should be disguised but rather a very important part of the research and the application of the results thereof.

On the other hand, one needs to be very careful in interpreting results obtained by the use of experienced panellists because they are generally far more critical than the average consumer. Yet, at the Danish Meat Products Laboratory, the experience is that a score of -1, - in the early days -2 was suggested but that seemed too lenient - by the panel is quite close to a point where a very discerning consumer may not be fully satisfied, at least after a period of home freezer storage of the product.

Coordinating taste panels

In much scientific work, and especially where international collaboration is considered, it is often proposed that one should come to an agreement with regard to scoring of various organoleptic attributes evaluated by taste or consumer panels. From what is indicated above, it is seen that this is unlikely to be possible, and at any rate it would lead to a peculiar situation where taste panel findings no longer necessarily corresponded to the prevailing expectations as regards food quality in the area where the research was carried out.

It would namely mean that one might have arrived at an agreed taste to be designated, for instance "acceptable", "neither good nor bad", or "average". If this were done with for instance tomatoes, a peculiar situation would exist. People in Northern Countries normally prefer tomatoes that are quite firm, almost greenish, etc. Someone from Italy or the Southern United States would consider these tomatoes unacceptable. They would demand a fully ripe, rather soft, aromatic, deep red fruit, which will actually be objectionable to a Northern palate. It is the firm conviction of the author that in the first place it would not be possible to educate taste panels to the extent that they could disassociate themselves completely from prevailing taste preferences in their daily life; besides, as mentioned, by doing so one would disguise important facts rather than illuminate anything.

Other factors than time and temperature

As described in considerable detail below, the shelf life of frozen foods depends not only on time and temperature of storage but also to a large extent on product, process and packaging, the so-called "PPP"-factors. Judgements as regards acceptability time (shelf life, PSL) depend also to a considerable extent on the panel used; therefore, it has been suggested by Löndahl and Danielson (1972) to include also the panel factor, i.e. to talk about TTT-PPP-P factors. This applies mainly where acceptability times are determined by a scoring method. If stability times are determined by a triangle method, the influence of the panel is negligible if the panels are experienced. The PPP-P terminology is not used in the present work.

One further word of caution is indicated. Taste panels are usually highly trained. They normally receive an introductory training, and their regular participation in panel sessions, scoring of products, etc., involves a certain discipline, which has to be acquired. This in itself results in the difficulty that the taste panels are often trained, at least subconsciously, to like the product which is the preferred prevailing product in that area at that moment. They may well react negatively to any product which differs from that standard, consider it defective and downgrade it, even if the general consumer might well accept and eventually even prefer it. As an illuminating example, had it been up to experienced coffee taste panels, instant coffee would never have been introduced on the market.

It is for that reason that food companies often carry out market tests using large number of samples and sophisticated preference testing systems among a stastistically selected sample of the potential consumers. However, even this is

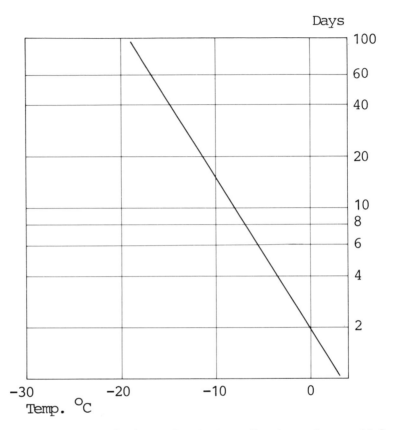

Fig. 44. Stability times, HQL, for raspberries in retail packages, frozen with 3 part berries to 1 part of a 50% sucrose syrup. Note the linearity of the curve even above the thawing point. After Guadagni, Nimmo and Jansen (1957c).

not too reliable since consumers appear whimsical. Also, of course, they are very influenced by price. They may register a preference for a product, which they would not get into the habit of buying. Therefore, commercial companies often resort to simply test marketing a new product in some areas to determine consistency of any preference.

Where taste panels are used, it is useful to keep these factors in mind and from time to time verify the panel's scoring with consumers' preference and, at times, include non-frozen and just frozen samples in the test.

Comprehensive test

Regardless of what test method is used, one must be aware of the fact that many factors influence consumers' choice and consumers' evaluation of a product. Thus, the taste panel in Albany found almost a straight line for various fruits frozen in sugar solution, e.g. as illustrated in Fig. 44.

These tests were carried out by presenting the thawed products, ready for consumption, to taste panels. However, in a sugar frozen product, partial thawing may have occurred in the products stored at comparatively warm temperatures, i.e. even at -10°C, cf. Table 10. Thawing a sugar frozen product normally results in an unsightly appearance of the package, syrup may even leak and damage other packages, etc. This suggests that in this case the organoleptic test might have included an examination of the appearence of the frozen product as it would be presented to the trade and to the consumer if the test were to represent the reaction of trade and consumers.

In other tests, e.g. those illustrated in Tables 5 and 6 and Fig. 8, no difference was recorded between quickly frozen and slowly frozen poultry. However, the quickly frozen products might, as mentioned above, have a very whitish appearance and be the one preferred by the consumer at the point of purchase. This difference would be important but remain unrecorded by a taste panel receiving cooked samples. Again, an evaluation of the frozen product might have been included in the test and have shed some additional light on the relative acceptability of the various products.

In general, the author recommends that more attention be paid to the choice of quality criteria in experiments related to the quality of frozen foods and that several different factors be considered.

Objective tests

Because of the variability of taste panels, it is often suggested that physical, chemical or other objective methods be used for quality evaluation of foods, and many sophisticated methods and instruments have been developed for this purpose. In fact, while one of the objections to taste panels is that they are somewhat laborious to work with, many of the instruments which have been proposed for their replacement are more time demanding and certainly more expensive in use than taste panels.

For one type of rather simple test, Winger (1982) found little correlation between TBA number and peroxide number and the quality of lamb.

However, and more important, the concept may be based on a fundamental misunderstanding which has been discussed by the author at considerable length in Jul and Zeuthen (1981). The purpose of practically all research on food quality is to determine the acceptability of the product to the consumer and the value which the consumer would attach to it. In fact, tea tasters in London score the samples, not by any sophisticated scoring system, but simply by indicating the price which they recommend as a buying price at next day's auction for the lot represented by each sample. Similarly, the purpose in devising any objective methods such as a determination of the degree of change of chlorophyll into pheophytin, content of trimethylamin, formation of hexanal, or the like, or the tear strength of muscle fibres, is that they should correlate with consumers' reactions. Therefore, any such method has to be carefully compared with consumers' acceptance characteristics. If consistently good correlation is found, they may be used as - hopefully - convenient replacements for organoleptic testing. However, wherever there is a disagreement, the consumers' acceptance will be the deciding factor where a quality judgment is to be made. This seems to be difficult to accept for many research workers. Further, many find it particularly disturbing that consumers' expectations vary over the years and from place to place. This means that objective test methods which have reflected consumers' acceptability fairly accurately at one time, may at other times have to be discarded, or at least the standards may have to be changed. It is quite obvious, however, that the objective of food research is to determine to what degree a product is acceptable to the consumer and cannot tell the consumer what she or he should like.

Changes in Nutritive Value

Effects of freezing on nutritive value

The above considerations have related to the influence of freezing on appearance, tenderness, taste, flavour, etc., and some other quality indices such as retention of ascorbic acid, breakdown of chlorophyll into pheophytin, etc. It is obvious that another useful criterion for the effect of freezing and freezer storage could be the influence of these treatments on the nutritive value of the product.

Research into nutritive value is somewhat different from other quality tests. Here the objective may be to determine the retention of nutrients, i.e. objectively measurable characteristics. However, even here, the biological availability rather than just the chemical presence of a factor has to be determined, ultimately with research involving animal experimentation, or, where ethically possible, even human subjects. Also, some subjective factor may have to be taken into consideration anyway, simply because if a consumer dislikes a product, it may not be eaten, regardless of its nutritive value. Conversely, if a product is improved in attractiveness, made cheaper or more easily available, the consumption of it may increase with the corresponding effect on the population's intake of nutrients.

Reference has been made above to protein denaturation during freezing, which is related to drip. Measurements of protein value have indicated that changes in this are so small that they do not result in any measurable reduction in the nutritive value of the protein.

Similarly, much reference has been given to lipid oxidation. However, human taste buds are very sensitive, and humans object to oxidation and rancidity long before these are of such magnitudes that they could have any measurable effect on the foods' nutritive value and wholesomeness.

The leaching out of dissolved material, which is characteristic for drip and

Table 26. Ascorbic acid in peas after various treatments. While freezing causes some losses, these seem to be compensated for by a correspondingly lower loss in final cooking before serving After Mary Morrison (1974).

	mg vit. C/ per kg dry weight
Commercially vined	123.2
Frozen	109.8
Frozen/cooked	75.7
Handpodded	139.4
Handpodded/cooked	76.9

for a somewhat increased cooking loss, is experienced with frozen foods, and can result in a loss of water soluble vitamins and minerals. Such measurements are generally included in measurements of the effect of freezing on nutritive value. The losses as recorded have mostly been small.

The most important adverse effect of freezing and frozen storage on nutritive value may be a loss of vitamins, mostly the more labile ones, such as ascorbic acid, thiamin, and riboflavin. One needs, of course, to view losses of nutrients in a realistic light. Thus, Mary Morrison has compared the losses of ascorbic acid in frozen peas, cf. Table 26. For similar conditions, Strachan (1983) reports even more favourable findings for frozen spinach, green beans and Brussels sprouts.

In this, one will notice that there is a considerable loss in home cooking of peas. In the conventionally frozen peas a significant loss is found during processing, primarily during blanching. The result is that when frozen peas are purchased, they have a lower vitamin C content than the unfrozen peas. However, this is offset by a smaller loss during home preparation. The result is that the product when eaten may have practically the same ascorbic acid content regardless of whether the product was frozen or not. In other cases, freezing may not rate this well. For instance, especially the blanching process prior to freezing vegetables often results in losses in ascorbic acid, which in total exceeds considerably those that occur in reasonable home cooking.

Fennema (1982) reviews some findings regarding nutrient losses due to freezing.

As a background for the possible nutritional consequences of freezing foods, an example of the effect of freezing in the meat industry and in consumers' homes may have on the nutritive value of the Danish diet is given below.

Meat's contribution to the nutritive value of diet

Table 27 shows the amount of the total intake of energy, protein and various vitamins and minerals in the average Danish diet which in each case is

Table 27. Average contribution of meat to nutrient supplies in Denmark, and USA, in per cent of total average intake. (- means no data available). (J. Hoejmark Jensen, 1981).

	Denmark 1973	USA 1974
Energy	15	19
Protein	24	39
Fat	28	34
Vitamin A	24	22
Thiamin	28	26
Riboflavin	26	24
Niacin	40	43
Pyridoxin	—	43
Cobolamin	—	66
Vitamin C	3	10
Iron	30	27
Calcium	1	3

supplied by meat. However, such a calculation may be somewhat misleading, i.e. it is seen that meat supplies 26% of the daily intake of riboflavin, which suggests that the intake from meat of this factor is quite important nutritionally. However, it appears as if the average diet contains very generous amounts of riboflavin, i.e. about 140% of the average requirement. Thus, even without the contribution from meat, the diet would be more than sufficient as regards this vitamin. In that case, even a considerable loss of riboflavin caused by freezing meat would have no real effect on the nutritive consequences of the diet. Table 28 shows that a corresponding situation exists for several other intakes.

A somewhat better indication of meat's role in the diet is obtained from Table 29. In this, the first column gives the content of the various nutrient factors in the average daily diet, calculated per 10 MJ, i.e. its density in the diet. When it is assumed that everybody in society except infants eats approximately the same diet, it can be calculated how much the diet should contain per 10 MJ of each nutrient factor to give an adequate intake of that particular factor, i.e. the recommended density of that factor for the group with the highest relative requirement. According to the Nordic nutritional recommendations, these should be as indicated in the second column of Table 29. The third column in that illustration indicates the percentage of the density of that nutrient in the food compared to the recommended allowance; the last column indicates the degree of that intake which is derived from the meat intake. It is worth stressing that nutrition recommendations generally incorporate a considerable margin of safety, because requirements differ from individual to individual as does food intake and food selection. If average

Table 28. The average intake of nutrients per capita per day in USA (1974) including and excluding the contribution from meat. For comparison the Recommended Dietary Allowances for adults are listed (NAS, 1980). After J. Hoejmark Jensen (1981).

		Average intake	Average intake excluding the contribution from meat	RDA for Males 23-25 years	RDA for Females 25-50 years
Energy	kcal	3350	2680	2700	2000
Protein	g	101	59	56	44
Fat	g	158	103		
Vitamin A	IU	8200	6400	3300	2700
Thiamin	mg	1.94	1.40	1.4	1.0
Riboflavin	mg	2.33	1.75	1.6	1.2
Niacin	mg	23.4	12.7	18	13
Pyridoxin	mg	2.28	1.23	2.2	2.0
Cobolamin	μg	9.7	2.9	3.0	3.0
Vitamin C	mg	119	107	60	60
Iron	mg	18	13	10	18
Calcium	mg	950	917	800	800

intake exceeds the recommendations, it is generally considered adequate for most population groups.

The role of meat in securing an adequately balanced diet may be estimated from the last two columns in Table 29, suggesting the following:

Protein. In general, the Danish population is quite well supplied with proteins. Meat's role in this supply is significant, and serious damage to the proteins caused by processing might have nutritional consequences. However, even if freezing can result in some moderate denaturation of proteins, availability and amino acid patterns do not seem to be measurably affected. Thus, changes in proteins due to the use of frozen meat is not a nutritional concern.

Calcium. The diet seems generally well supplied with calcium, and the contribution from meat is so small that any loss of calcium during meat freezing and thawing would not be of nutritional significance.

Iron. The diet seems undersupplied with this mineral, so much so that one may somewhat question the recommended requirements since there is little evidence of widespread anaemia or similar occurrences. Nevertheless, it is known that anaemic diseases are quite frequent in Western societies. About half of the iron intake is derived from meat. Therefore, any loss of this mineral

Table 29. Nutrient content of the average Danish diet, from Helms (1981) and Helms (1978), highest relative requirement from Statens Levnedsmiddelinstitut (1981), and degree of sufficiency of diet and contribution thereto from meat and fish. Data also from J. Hoejmark Jensen (1981).

	Content in diet per 10 MJ	Requirement per 10 MJ for group with highest relative requirement	Degree of suf- fiency %	Part of suf- fiency derived from meat %
Protein, g	71	60	117	24
Retinol. mg	1.62	1.0	162	24
Thiamin, mg	1.90	1.3	140	28
Riboflavin, mg	2.13	1.5	142	26
Pyridoxin, mg	2.3*	2.5	92	39*
Niacin equivalents, mg	24	16	150	40
Cobolamin, mg	7.1	3.8	187	66*
Calcium, g	1.1	0.9	120	1
Iron, mg	11.6	22	53	30

* US data, Spring 1977 (USDA, 1980), include fish.

due to freezing and thawing would have serious consequences. In addition, the iron in meat is known to be biologically easily available, a factor which further stresses the importance of avoiding any losses of this mineral during freezing and thawing meat. However, since freezing does not affect iron availability and drip from meat normally is limited to a few per cent, freezing of meat is unlikely to have any significant consequences in so far as iron is concerned.

Retinol. The Danish population seems well supplied with this vitamin and even considerable damage, which might be caused by freezing, appears not be of nutritional significance.

Thiamin. The average diet contains an ample amount of thiamin so even a considerable loss of that part of the thiamin which is derived from meat would hardly have any nutritional significance.

Riboflavin. The situation for this vitamin is much the same as for thiamin.

Pyrodoxin. The Danish diet seems already only modestly supplied with this vitamin. Especially among the elderly (Elsborg, 1962), there may be symptoms of pyridoxin deficiency. Also, the use of oral contraception may result in an increased requirement. Further, Ellis *et al.* (1980) have demonstrated a highly beneficial effect of pyridoxin in the prevention or cure of the carpal tunnel syndrome. Such studies further emphasize the justification for special attention to the diet's pyridoxin content. About 40% of the pyridoxin intake is derived from meat. Thus, any loss caused by processing could have substantial, adverse nutritional significance. It is, however, worth stressing that pyridoxin analyses cause difficulties. Thus, intake is not known with any high degree of accuracy. Also, doubts exist with regard to requirements. Nevertheless, the effect of freezing meat on this vitamin is worth close scrutiny.

Niacin. The diet in Denmark seems amply supplied with niacin. A significant amount is derived from meat, but even a substantial loss of the contribution of niacin from meat which might conceivably be the result of freezing meat would not be likely to lead to any deficiency situation.

Cobolamin. The average diet seems very well supplied with this vitamin. As mentioned above, even though a major proportion, namely 66% of requirement, is supplied by meat, a substantial loss of this part of the intake would not render the diet deficient with regard to this vitamin. A concern is often expressed that a diet without animal products may easily be deficient in vitamin B_{12}. However, Immerman (1981) finds this to be a rare occurrence, even for vegans, i.e. complete vegetarians, not eating eggs nor dairy foods.

Ascorbic acid. Only 3% of this vitamin is derived from meat. Therefore, it is hardly meaningful to discuss consequences on the diet of a loss in this factor due to the freezing of meat.

It is clear from the above that for meat, any nutritional concern as regards freezing of meat must consider mainly the effect on vitamins of the B-group.

Meat intake in Denmark

The most recent data for the consumption of meat in Denmark are found in Danmarks Statistik (1981) and reproduced in Table 30.

B-vitamins in meat

In calculating the contribution of this intake to vitamin intake, one has to know the vitamin content of the meat, of course. Here, it is worth noting that there are quite considerable differences in the vitamin content of the various cuts of meats, mainly depending on the fat content of the cut, since increasing

Table 30. Consumption of meat in Denmark, calculated after Danmarks Statistik (1981).

	Consumption per cap. per day in g		
	1978	1979	1980
Pork	137.5	144.1	149.2
Beef and veal	52.3	46.6	44.4

Table 31. Thiamin and riboflavin content in cuts of pork in Denmark, in μg per 100 g, calculated after Hansen *et al.* (1976).

Fat, %	Designation	Thiamin	Riboflavin
less than 5	Trimmed meat	1070	310
approx. 10	Lean meat	770	220
approx. 20	Medium fat meat	710	220
approx. 30	Fatty meat	590	190
approx. 40	Lean fatty tissue	480	160
approx. 50	Fatty tissue	230	150

Table 32. Content of B vitamins in 100 g pork. After Helms (1978).

	Pork			
	Trimmed	Lean	Medium fat	Fat
Fat, %	3	10	20	30
Niacin equivalents, mg	7.9	7.7	6.4	5.6
Thiamin, μg	1070	810	810	530
Riboflavin, μg	310	195	210	220
Pyridoxin, μg	500	400	330	350
Cobalamin, μg	5	1	1	1
Folates (total), μg	9	7	6	9.1
Pantothenic acid, μg	700	700	680	680
Biotin, μg	5	5	5.1	5.1

fat content generally means reduced vitamin B content as shown in Table 31.

The data in this illustration are average figures for pork from a rather detailed study carried out by the Danish Meat Research Institute and the Danish State Food Institute. It included only thiamin and riboflavin.

Helms (1978) gives the figures for contents of all eight B-vitamins in meat as indicated in Table 32.

Table 33. B-vitamin content per 100 g in beef and veal. After Helms (1978).

		Beef		Veal
	Lean	Medium fat	Fat	
Fat, %	3	7.5	24.5	2-11
Niacin equivalents, mg	10.5	9.5	8.2	10.8
Thiamin, μg	95	85	75	125
Riboflavin, μg	190	175	150	170
Pyridoxin, μg	500	450	380	400
Cobalamin, μg	2	2.2	1	1.2
Folates (total), μg	7	7	7	5
Pantothenic acid, μg	1000	600	600	900
Biotin, μg	4.6	3.8	—	—

Table 34. B-vitamin content per 100 g in beef and veal. After Ege (1978).

	Beef			Veal		
	Lean	Medium fat	Fat	Lean	Medium fat	Fat
Fat %	13	19	25	8	13.5	17
Niacin equivalents, mg	9.5	9.0	8.0	10	9.5	9.0
Thiamin, μg	80	70	60	120	100	90
Riboflavin, μg	180	165	150	250	225	200

Table 35. Thiamin and riboflavin content of beef and veal, calculated after Hjarde *et al.* (1981), by Lüthje (1982).

Fat content	Thiamin μg/100 g	Riboflavin μg/100 g
Less than 5%	58	189
5-10%	50	164
approx. 15%	63	214
approx. 25%	36	136

The consumption of pork, which is given in Table 30, is an average figure calculated on the basis of total disappearance in a year. Since a hog carcass without kidney fat contains about 30% fat, it may be assumed that the pork contains an average of about 30% fat. In the following calculations, therefore, data for meat with 30% fat have been used for pork. The contents of thiamin and riboflavin are calculated in accordance with the data in Table 31, the other vitamins in accordance with Table 32.

For beef and veal, Helms (1978) gives the vitamin B contents quoted in Table 33. In this the same vitamin content is assumed for lean, average and fatty veal. It is likely that this is due to insufficient data.

Ege (1978) gives the data found in Table 34. Table 33-35 indicate that the thiamin content in beef and veal is considerably lower than in pork. Conversely, veal and beef contain slightly more niacin than pork. The content of other B-vitamins is approximately the same in these three types of meat. For beef and veal it is somewhat difficult to determine which average vitamin content should be used in calculations. Firstly, there are differences between beef and veal, and secondly, beef is an even less well-defined product than pork. Further, Hjarde *et al.* (1981) maintain that there is no clear relation between the fat content of beef and its content of riboflavin, e.g. the tenderloin contains about 245 mg riboflavin per 100 g, while the content in other cuts with the same fat content is only about 175 mg per 100 g.

Hjarde *et al.* (1981) carried out a detailed study of the nutritive value of lean, normal, and fat beef cattle and veal, and analyzed a large number of different cuts. Lüthje (1981) grouped the cuts according to their fat content and suggested the vitamin content as indicated in Table 35.

In this, thiamin content is found to be only half of what is suggested in Table 33 and 34, while there is a general agreement about the content of riboflavin.

Table 36. The contribution to daily requirements of B vitamins from pork, beef and veal in the average Danish diet, 1979. (Lüthje, 1982).

Vitamin	Daily require- ment	Daily from: pork	intake beef and veal	Percentage of daily requirement from: pork	beef and veal	Total meat
Thiamin, μg	1500	850	23	57	1.5	58
Riboflavin, μg	1700	274	82	16	4.8	21
Pyridoxin, μg	2200	504	210	23	9.5	32
Niacin- equivalents, mg	19	8.1	4.4	43	23	67
Cobolamin, μg	*3.0	1.44	1.03	48	34	82
Folates (total), μg	*400	13.1	3.3	3.3	0.8	4.1
Pantothenic acid, μg	*600	980	280	163	47	210
Biotin, μg	*150	7.3	1.8	4.9	1.2	6

* Data from USA. Other requirement data from Statens Levnedsmiddelinstitut (1981).

In the calculations below of the contribution of beef and veal in the diet towards intake of thiamin and riboflavin, an average of the thiamin and riboflavin content given by Hjarde *et al.* (1981) is used, i.e. 50 microgram thiamin per 100 grams and 125 microgram riboflavin per 100 g. For the other B-vitamins, the data from Table 33 have been used.

Meat's contribution to B-vitamins in the Danish diet

On the basis of the above data and the daily consumption of pork, beef and veal, given in Table 30, the average daily intake of B-vitamins from meat has been calculated and is reproduced in Table 36.

In considering this, one must keep in mind that only fairly rough estimates of vitamin contents were available and that no allowance has been made for losses in processing, storage, distribution, preparation in the home, and - in certain cases - warmholding. It is seen that the largest contribution comes from pork, mainly because of the high intake of pork as compared to beef and veal.

Except for biotin and folates, meat covers a significant amount of the daily intake of B-vitamins. The highest relative contribution is pantothenic acid where the daily meat intake covers twice the recommended daily intake. For niacin, thiamin, and cobolamin, meat covers about half of the recommended intake; for riboflavin it covers about 1/5th. In so far as the four latter vitamins are concerned, however, they are already well-covered in the Danish average diet as mentioned above and as indicated in Table 29. Thus, any loss of these vitamins due to freezing of meat is unlikely to have any significant consequence on nutrition in Denmark.

Pantothenic acid seems widely distributed in foodstuffs, and deficiency symptoms among humans are rare. Although there is some evidence that the pantothenic acid content in meat may be reduced considerably by freezing, this is hardly a genuine nutritional concern. Yet, it seems relevant to seek further data regarding the extent of such losses.

For pyridoxin, however, Table 29 shows that intake is only 92% of recommended intake. About 1/3rd of this seems to be derived from meat. This suggests that the retention of pyridoxin in meat is of special significance.

Intake of frozen meat

The significance of the influence of freezing on retention of vitamins of the B-group in meat has to be judged in relation to the total intake of frozen meat.

Excluding frozen ready-to-eat dishes, the total consumption of frozen meat in Denmark amounted to 14 500 metric tons in 1980 (Statistisk Aarbog, 1981). The composition of this intake is given in Table 37.

Table 37. Composition of average Danish frozen meat purchases in retail shops. After Dybfrostinstituttet (1981).

Ground	16 %
Ground, shaped	19 %
Whole cuts	19 %
Danish pork sausage	13 %
Other sausages	14 %
Cold cuts	9 %
Offal	2 %
Other meat	10 %

Table 38. Intake of frozen prepared meat foods in Denmark, after Dybfrostinstituttet (1981).

Meat patties, fried	7 %
Meat dishes with gravy	2 %
Meat and flour balls	39 %
Spring rolls	13 %
Soups	26 %
Pizzas	5 %
Others	10 %

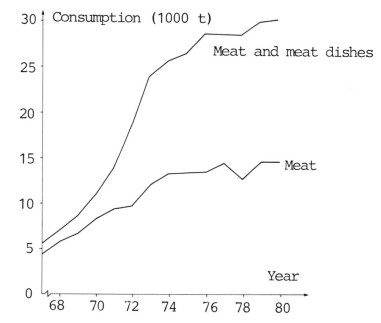

Fig. 45. Trend in the use of frozen meat and meat dishes in Denmark. After Dybfrostinstituttet (1981).

Dybfrostinstituttet (1981) gives the intake of frozen ready-to-heat dishes which was of about the same magnitude. Out of the latter, only 60% were meat dishes. It is assumed that the meat content of these is 50%. Thus, the contribution from ready-to-eat frozen dishes to frozen meat intake will be about 4 400 tons. The composition of this intake is reviewed in Table 38.

Altogether, consumption of frozen meat, purchased as such, or as frozen prepared dishes, was about 19 000 tons in 1980 or 3.5 kg per capita per year. According to Statistisk Aarbog (1981), total meat consumption was 69.6 kg per capita in 1981. Thus, 3.5 kg is equal to about 5.0% of the total meat consumption. This is the direct intake of frozen meat purchased in retail establishments. The intake seems to have stabilized at this level, cf. Fig. 45.

Freezing meat for further processing

Retention of nutrients during freezing and freezer storage is also of importance in cases where meat is frozen as a raw material, i.e. as whole or split carcasses, quarters or vacuum packaged primal cuts, etc., for use later in meat processing, e.g. canning, curing, or the preparation of frozen products; no overall data exist with regard to the extent of the use of this technology. Therefore, attempts were made to obtain data regarding this practice. For one large Danish retail chain, practices varied widely with regard to different meat groups and different seasons. However, Lüthje (1982) indicates that the data in Table 39 may be indicative of the practices of that retail chain.

As mentioned above, these data have been obtained from one chain only. It is likely that practices in various establishments vary considerably. It may be interesting, however, to calculate on the basis of the practices indicated in Table 39 the amount of frozen meat which would go into the retail trade through such products if the same practices were followed throughout the country.

The total consumption of meat products in Denmark was 79 320 tons in 1979 or 15.6 kg per person. Out of this, about 1 kg per person is sold frozen, and thus is included above. The remaining 14.6 kg per capita is not sold frozen. If one assumes that the use of frozen raw material is as indicated in Table 40, one may estimate the total amount of frozen, meat which goes into meat products, not sold frozen to 25% of 14.6 kg or 3.7 kg per person per year, or 5.3% of daily meat consumption per person.

Additionally, frozen meat is often used thawed and sold as fresh meat. The chain mentioned above indicates the degree to which frozen meats are used in these as indicated in Table 41 (Lüthje, 1981).

Again, no data exist which indicate to what degree these data are applicable for Danish meat consumption in general. Especially if the data for ground meat are indicative, this would represent quite a significant part of the Danish

Table 39. Estimated extent of use of frozen meat in a supermarket chain in Denmark. (Lüthje, 1982).

Beef for curing and roast beef Unfrozen, when available, most frequently supplemented with frozen	50 %
Pork bellies for rolled sausage Practically always frozen	100 %
Sausages, frankfurters, etc. Frozen meat returned from stores, raw material often stored frozen	25-50 %
Prepared dishes Mostly frozen meat raw material	90 %
Smoked pork loins At least half are frozen	50 %
Excess products Whenever raw materials have been purchased in excess of demand because of changing market conditins, it is frozen	15 %

Table 40. Assumed use of frozen raw material in various meats. After Lüthje (1982).

50 % in cured beef
80 % in rolled sausage
25 % in other sausages
50 % in smoked pork loin
10 % in cured pork loin

Table 41. One Danish supermarket chain's use of frozen raw material for various fresh meat products. After Lüthje (1982).

Ground meat (not tartar)	25%
T-bone steaks, osso-buco	estimated 15%
Mutton and lamb	95%

meat consumption.

If it is assumed that total consumption of steaks and chopped meat is 40% of total fresh meat consumption less processed meats, which is 69.6 less 15.6 or 54 kg, and that 15% thereof is prepared on the basis of frozen raw material, this suggests that 4.7% of the Danish fresh meat consumption of meat not sold frozen has been frozen through such practices prior to sale, i.e. 3.6% total meat consumption.

Finally, one must consider freezing in the homes. There are home freezers in about 70% of Danish homes as indicated in Fig. 1. Many of these will be comparatively small freezer compartments in home refrigerators. A not insignificant number, however, will be home freezers of the trough type or separate vertical freezer cabinets. Especially in rural areas, where freezing of home produced products is common, fairly large amounts, especially of pork, will be preserved in that form. It may be assumed that home freezers on the average will contain about 4 kg of meat, not purchased frozen, at any one time. An average storage period of about 4 months may be estimated. Thus, 12 kg or 17.2% of the population's annual meat consumption not previously frozen will have passed through a home freezer.

As indicated above, some frozen meat is being used after thawing for preparation of retail packaged frozen meats or frozen meat dishes. Similarly, some products, which are sold fresh, have been prepared from frozen materials, and may be frozen again in home freezers. Thus, a certain amount of repeated freezing will take place and needs be taken into account when the total effect of freezing meat on nutrient intake is considered. Because of insufficient data, this consideration has not been included in this calculation.

When adding up the totality of meat in the Danish diet affected by freezing, one thus arrives at a figure of about 31%, with about half of that frozen in the homes.

Institutional meat supply

Freezing may attain special significance in the area of institutional catering and the supply of meals from institutions to private households. In this regard, freezing seems to be utilized more and more. In the past, many of these food supply systems were based on warmholding the food. The meat used in this way has often been frozen. Additionally, distributing the food in the frozen form may have many advantages, especially where the distribution of food to private households is concerned, and it may be gaining in popularity. For one thing, it is convenient for the receiver of such meals to be free to determine when the meal will be eaten. Also, in this way the user is assured that the food is adequately hot when served. It does appear as if both food quality and nutritive value may be better retained in such a cook-freeze system than in the previously used systems. Also economically, it does appear as if a cook-freeze-heat system is more advantageous since productive capacity and personnel can be utilized more effectively and distribution can be more economic (Bech-Jacobsen and Klinte, 1981a).

While no detailed data exist about the extent of the use of freezing at present in these systems, it is known that it is used widely in the UK for school feeding and in Sweden and Norway for feeding the elderly. It may be of importance for that section of the population which is served by such systems and the consequences nutritionally must be considered in their introduction.

Retention of B-vitamins in frozen meat

It is seen from the above that a considerable part of the meat consumed in Denmark has been frozen at some point. In addition, it seems that this percentage is increasing. Therefore, the nutritional effect of meat freezing is in need of consideration. As mentioned above, here mainly the effect on the supply of B-vitamins in the diet need be considered. As above, the chemical designation for the main component in each vitamin group has been used, e.g. cobolamin instead of vitamin B_{12}.

In most cases, it is somewhat unclear what has actually been determined in the various vitamin assays to which reference is made.

Considerable data are found in the literature as regards the influence of freezing on meat's content of B-vitamins. However, many of these data seem contradictory or conflicting. The reason for the latter is that many sources of error exist, e.g. very considerable differences between the results of various methods of analysis. Also, different contents are found in different animals, in different muscles of the same animal, and even between various cuts of the same muscle. Thus, differences in content may be due to differences between samples and controls. In spite of these many sources of error, it seems possible to make some general conclusions on the basis of existing data.

Losses of B-vitamins may occur during ageing of the meat prior to the freezing process itself, the freezing process, freezer storage, thawing or be due to drip, and losses during final food preparation and warmholding. Peculiarly enough, this subject was extensively studied in the 1940's and the 1950's, but few data exist from later years. Therefore, many recent review articles such as Engler and Bowers (1976), Fennema (1975), and Cutting and Hollingsworth (1974) are basically reviews of earlier works. This is unfortunate, because in the early years, considerable differences were found between different methods of analyses.

Influence of ageing

Westerman *et al.* (1955) determined the influence of various methods of ageing prior to freezing on the retention of thiamin, riboflavin, pantothenic acid and niacin in pork. Meat from *M.longissimus dorsi* was aged 1, 3 and 7 days at -1°C and 7 days at +4°C and was subsequently stored for 48 weeks at -18°C.

Thiamin retention was as low as 66% after 24 weeks of storage and 79% after 48 weeks when the samples had been aged for 3 days. Retention of riboflavin was 81% after 24 weeks and 86% after 48 weeks.

None of the other conditions gave significant losses. For pantothenic acid and niacin, lowest retentions were found in the samples ripened one day at

Table 42. Retention of B-vitamin after freezing and thawing, without freezer storage. Compiled by Lüthje (1982).

		% Retention					
	Temp. (°C)	Thiamin	Ribo-flavin acid	Niacin	Panto-thenic	Pyri-doxin	Reference
Pig, *Long.dorsi*	−18	83	76	94	82	78	Lee *et al.*, 1954
Pig, *Long.dorsi*	−18	79	117	78			Lehrer *et al.*, 1951
Beef, *Long.dorsi*	−18	94	96	93	102	98	Lee *et al.*, 1950

-1°C. The retention of pantothenic acid was 88% after 24 as well as 48 weeks; the losses of niacin were small and not significant.

Ripening of *longissimus dorsi* and *semimembranosus* for 21 days at 1°C prior to freezer storage for three years at -18°C had no effect on riboflavin retention. On the other hand, ageing reduced thiamin content and increased niacin content compared to the unripened control according to Meyer *et al.* (1963). In conclusion, ageing prior to freezing may result in some losses of vitamins in the B-group, especially of thiamin and riboflavin. Such losses are probably included in most of the freezing and storage experiments quoted below and probably need not be considered separately.

Influence of freezing

Table 42 gives the results of three studies on the effect of freezing alone. No freezer storage period was included. The effect of freezing alone was determined in all three cases in meat from *longissimus dorsi* which was frozen to a temperature of -18°C and thawed again within 24 hours.

For pork, Lee *et al.* (1954) found a retention of thiamin, riboflavin, pantothenic acid and pyridoxin around 80% and for niacin 94%.

Lehrer *et al.* (1951) found about the same figures for thiamin but a lower retention for niacin. The largest difference was found for riboflavin with retention of 76% and 117% respectively. Freezing and thawing of beef resulted in only small changes in the content of thiamin, riboflavin and niacin according to Lee *et al.* (1950), as shown in Table 42.

Table 43. Retention of B-vitamins after freezer storage. Retention compared with content in a sample frozen and thawed within 24 hours. Compiled by Lüthje (1982).

	Time (Days)	Temp. (°C)	Thiamin	Ribo-flavin	Niacin	Panto-thenic acid	Pyri-doxin	Reference
					% Retention			
Pig, *Long.dorsi*	168	−18	85	94	95			Westerman *et al.*,1955
Pig, *Long.dorsi*	180	−18	87	56	115			Lehrer *et al.*, 1951
Pig, *Long.dorsi*	180	−18	121	106	112	103	82	Lee *et al.*, 1954
Beef, *Long.dorsi*	180	−18	96	100	112	89	78	Lee *et al.*, 1950
Beef, *Long.dorsi*	300	−18	106	57	99			Lee *et al.*, 1950
Beef liver	270	−20					113	Richardson *et al.*, 1961
Beef liver	450	−20					82	Richardson *et al.*, 1961

Rate of freezing

The three studies just mentioned included the effect of freezing rate. Lee *et al.* (1954) and Lee *et al.* (1950) used three different rates resulting in freezing times of 1, 5 and 19 hours respectively. Lehrer *et al.* (1951) used -18°C and -26°C for freezing. No significant differences could be determined as a result of freezing rate in the vitamin B retention due to freezing after 6 months' storage for pork and 10 months' storage for beef.

Contrary to this, Nestorov and Kazhukazowa (1975) conclude that rapid freezing (sharp freezing at -38°C for two hours) is necessary for a satisfactory retention of B-vitamins in meat and liver to be freezer stored for longer periods.

Influence of storage

Three studies have determined how the vitamin B content in pork is affected by a storage period. Comparisons were made between the vitamin contents in samples frozen and stored at -18°C with the content in samples frozen and thawed within 24 hours. The data obtained are given in Table 43.

It seems other factors have interfered. Thus, for pork, thiamin retentions between 85% and 121% were found. For riboflavin, Lehrer *et al.* (1951) found 56% retention while Lee *et al.* (1954) found 106% retention, i.e. a slight increase. For niacin and pantothenic acid no significant losses were reported, for niacin even a slight increase was reported. Freezer storage of beef for 300 days did not result in any reduction in the thiamin content. Riboflavin content was unchanged after 180 days, but reduced considerably with a retention of 57% after 300 days. Lowest retention of the other vitamins was 89% for pantothenic acid and 78% for pyridoxin, cf. Table 43.

Accumulated retention

An attempt has been made to measure total losses during freezing and freezer storage for pork. Data found in the literature are given in Table 44.

Some difficulties exist in consolidating these data because different storage temperatures and times were used. Further, there were differences in preparation of samples, packaging and freezing method, thawing, etc. In Table 44, data for pork are arranged according to warmer and warmer storage temperature, and longer and longer storage periods. Thus, the last data in the table are assumed to represent those samples which have suffered the most significant losses.

Best studied is thiamin. The table suggests retentions from 68% to 101%.

Quite a few data exist also for riboflavin and niacin retention. The data for riboflavin retention vary considerably, i.e. from 66% to 144%. Peculiarly enough, the sample which was stored for the longest period showed a large increase in riboflavin content. The variation is less for niacin with retentions from 90% to 114% suggesting that this vitamin is not significantly affected by freezing.

Retentions of 84% and 106% are found for pantothenic acid, while one result suggests a retention of 64% for pyridoxin.

The data obtained by Lehrer *et al.* (1951) were obtained from *longissimus dorsi* from 6 pigs. The analytical procedure was such that the vitamin content in each cut was followed, and retention was calculated for each animal. The very low retentions of thiamin and riboflavin, 68% and 66% respectively, cannot be explained by variations between animals. The data themselves are statistically significant, but great differences exist between retention in different animals. Thus, for samples stored at -26°C, retention in the six animals varied between 48% and 115% and at -18°C between 52% and 82%. The authors suggest different factors which may have caused the large variations. Firstly, the analyses themselves may have been subject to variations. Secondly, genetic factors may result in differences in vitamin

Table 44. Vitamin retention after freezing and freezer storage of pork. Compiled by Lüthje (1982).

	Time (Days)	Temp. (°C)	Thiamin	Ribo-flavin	Niacin	Panto-thenic acid	Pyri-doxin	Ascor-bic acid	Reti-nol	Reference
					% Retention					
M. Long.d.	180	−26	83	97	98					Lehrer et al., 1951
M. Glut.m.	90	−20	93						92	Nestorov and Kozhu-harova, 1975
M. Long.d.	90	−20	92						92	Nestorov and Kozhu-harova, 1975
M. Long.d.	180	−18	68	66	90					Lehrer et al., 1951
M. Long.d.	180	−18	101	81	106	84	64			Lee et al., 1954
M. Long.d.	365	−18	89	144	114	106				Wester-man et al., 1952
M. Long.d.	120	−16	78							Kemp, 1976
M. Biceps.	180	−16	81	71				80		Shestakov, 1976
Flank and loin	180	−13	92	100	98					Hartzler, 1949
Liver	90	−20	91						92	Nestorov and Kozhu-harova, 1975
Liver	180	−13	95	96	98					Hartzler, 1949

retention. Finally, there may be significant differences in the vitamin contents in the different parts of the same *longissimus dorsi*.

In a similar experiment, Lee *et al.* (1954) found generally better retentions of thiamin and riboflavin. Since, as indicated, rate of freezing does not appear to cause differences and since storage temperatures and storage times were the same, these factors cannot account for the differences. The only difference was

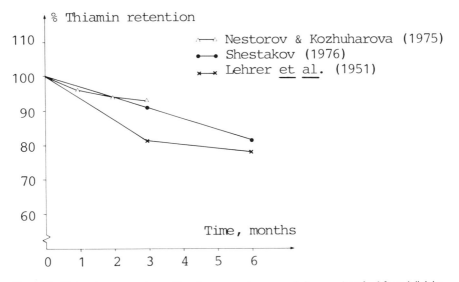

Fig. 46. Thiamin retention after freezer storage of frozen beef. After Lüthje (1982).

that Lehrer *et al.* (1951) thawed at ambient temperature, while Lee *et al.* (1954) prepared samples for analysis directly from the frozen meat. Table 44 indicates that Lehrer *et al.* found considerably lower thiamin retention than Lee *et al.* (68% against 101%), and also a lower retention of riboflavin (66% against 81%). No clear explanation seems to exist between the differences in the data reported.

An article by Westerman *et al.* (1952) was not available. The results given in Tables 43 and 44 are quoted from Fennema (1975). The retention of thiamin 89% and niacin 114% seems in accordance with most findings, while a riboflavin retention of 144% is considerably higher than reported elsewhere.

Nestorov and Kazhukazowa (1975) found low but significant losses of thiamin during storage for 6 months of *longissimus dorsi, glut. medius,* and pig liver. The authors attributed the quite favourable vitamin retention to rapid freezing.

A study by Kemp *et al.* (1976) was primarily designed to compare normal and PSE pork including thiamin content. They found that the thiamin content was significantly higher in normal pork than in PSE meat. However, they did not find any difference in retention between the two qualities of meat. Their studies resulted in a retention of 78% in both cases.

In the studies quoted above, storage temperatures from -13°C to -26°C were used. It is difficult to determine the influence of storage temperatures since other factors are also different between the different studies. In studies by Lehrer *et al.* (1951), storage at two temperatures were used while all other

Table 45. B-vitamin retention in beef after freezing. Compiled by Lüthje (1982).

| | Time Days | Temp. (°C) | % Retention | | | | | Reference |
			Thiamin	Ribo-flavin	Niacin	Panto-thenic acid	Pyri-doxin	
Ground thigh meat	120	−18	98	102				Clark and Van Duyne, 1949
Ground thigh meat	240	−18	98	102				Clark and Van Duyne, 1949
Long.dorsi	180	−18	92	91	101	91	76	Lee *et al.*, 1950
Long.dorsi	300	−18	98	57	96			Lee *et al.*, 1950
Ground *Long.d.* and *Semimem-branosus* Unaged	1095	−18	134	109	80			Meyer *et al.*, 1963
Ground *Long.d.* and *Semimem-branosus* Aged 21 days	1095	−18	97	108	101			Meyer *et al.*, 1963
Sliced beef liver	60	−18	73	66	110			Kotschevar, 1955
Chopped beef liver	60	−18	85	87	99			Kotschevar, 1955

parameters were the same. No significant difference was found between retention of thiamin, riboflavin, and niacin at -18°C compared to -26°C.

Generally, no discernable difference seems to exist between retention at different freezer storage temperatures. Some data suggest that the highest storage temperature investigated, -13°C, gave the best retention, about 100% for the three B-vitamins analysed, cf. Hartzler (1949).

In other works, however, there is a trend towards lower thiamin retention after increased storage periods, cf. Fig. 46.

Existing data for retention of vitamin B after freezing of beef and beef liver are summarized in Table 45. As is the case for pork, retention of thiamin, riboflavin, and niacin are most frequently determined. Actually, only Lee *et al.* (1950) appear to have determined retention of other B vitamins.

Thiamin retention of about 100% was found in practically all cases. In one case, Meyer *et al.* (1963) found a significant increase, a retention of 134%. In a test on beef liver, a lower retention, 73% to 85%, was found.

Retention of riboflavin is also generally around 100%. In a few cases, a slight increase is determined. In one case a retention of only 57% was reported.

The data for riboflavin retention in beef liver are somewhat lower than the retention in the meat.

Retention for niacin is also typically around 100%. In one case, a retention of only 80% is reported. The retention in beef liver is about as good as that in the meat itself. A study by Meyer *et al.* (1963) is interesting in that the meat was stored for as long as 3 years. The experiment included 8 animals. When the meat was aged for 21 days before freezing, there were no significant changes in thiamin, riboflavin or niacin content, even after this very long storage period. For unripened samples, there was no significant change in the content of riboflavin, a significant increase in the thiamin content, i.e. 134% retention, and a significant loss in niacin, (80% retention).

Lee *et al.* (1950) found a retention of 75% for pyridoxin.

Reference is made also to data found by Moleeratanond *et al.* (1981), showing a retention of thiamin, riboflavin and niacin in boxed beef which over a year was practically unaffected by whether the storage temperature was -18°C, -23°C or fluctuating between these temperatures. Thiamin retention was about 50% after 12 months' storage, with considerable differences between different products. Retention decreased almost linearly with time, independently of temperature.

Nutrients lost with drip

Since the B-vitamins are water soluble, some losses occur in connection with drip. Pearson *et al.* (1951) studied the vitamin loss with drip from pork and beef. In both studies, slices of *longissimus dorsi* were used. The data are given in Table 46.

For pork, the vitamin losses varied from 4% for riboflavin to 11% for niacin. Losses were in all cases larger for beef than for pork. This tendency was most pronounced for pantothenic acid.

The amount of drip from beef varied from 6.4 to 12.4% in Table 3 and thus was unusually high, cf. Table 12, For beef, Pearson *et al.* (1951) did not report the amount of drip.

Kotschevar (1955) found significantly lower amounts of drip from pork, 1.4 to 2.0%, and therefore, of course, also significantly lower losses of vitamins as given in Table 47. Kotschevar stored the meat for 60 days, whereupon it was thawed either 4 hours at room temperature or 20 hours at 4°C.

Pearson *et al.* (1951) did not include any storage time; thawing was carried out at 2°C for 18-20 hours. These differences in thawing conditions and storage time may account for the differences between the amounts of drip recorded.

Table 46. Loss of B-vitamin with drip. After Pearson *et al.* (1951).

	Loss of B-vitamin as percentage of total	
	Pork	Beef
Thiamin	9.0	12.2
Riboflavin	4.2	10.3
Niacin	10.7	14.5
Pyridoxin	8.7	9.4
Pantothenic acid	7.0	33.2
Cobolamin	5.1	—
Folic acid	—	8.1

Table 47. Per cent drip and loss of B-vitamins in per cent of original content. After Kotschevar (1955).

	Cut	% drip	Thiamin	Ribo-flavin	Niacin
				Loss with drip	
Pork	Tenderloin	1.4	0.90	0.70	0.72
	Loin	1.9	2.1	2.9	1.5
	Chop	2.0	2.4	3.0	1.5
Beef	Flank	2.2	5.1	0.4	3.1
	Rib steak*	6.3	16.8	6.3	1.5
	Rib steak	5.7	14.3	8.6	9.0
	Rib roast	3.2	5.3	3.2	5.3

* Thawed at ambient temperature. All others thawed in refrigerator.

In the experiment with beef, Pearson *et al.* (1951) used a higher thawing temperature, namely 26°C for 14–15 hours, resulting in somewhat more rapid thawing than that which was recorded by Kotschevar *et al.* (1955). The latter's experiment showed for most cuts only very minor vitamin B losses with drip and the authors concluded that it is doubtful if loss of nutrients with drip is large enough to be of any significance nutritionally. Pearson *et al.* (1951), on the other hand, felt that the considerable drip losses from beef, which they found, justified the collection and the use in food preparation of the drip.

Fennema *et al.* (1975) found that the amount of drip varies according to the product type, natural variations within the product, and the process, e.g. freezing and thawing, to which it has been exposed. Drip from pork and beef may vary between 1 and 10%; pH influences drip considerably, a high ultimate pH resulting in lower drip. For beef, they concluded that ageing and rapid freezing result in the smallest amount of drip.

Table 48. Average vitamin retention in fresh pork loin and ham after cooking not using drip in food preparation. After McIntire *et al.* (1943).

Preparation	% Retention		
	Thiamin	Ribo-flavin	Niacin
Roasting	60	85	80
Braising	70	75	65
Broiling	70	75	85

Vitamin B retention during kitchen preparation

Practically all meat is cooked, i.e. undergoes some kind of heat treatment before it is eaten. This will result in a certain loss of B-vitamins. The amount of loss depends on the method of preparation. Table 48, which gives retention after various preparation methods, indicates that thiamin, which is the most thermolabile of the B-vitamins, also shows the greatest loss whether the meat is roasted, braised or grilled. Braising resulted in lowest retentions. It is suggested that this is because water, which is added during braising, may result in a certain leaching out of vitamins.

If the drip water is utilized in the cooking and thus included in the finished dish, the same retention was found for all three preparation methods, on the average 76% for thiamin and 85% for riboflavin and niacin (McIntire, 1943).

The heat treatment of pork loin and ham is comparatively brief. Beef kidney and pig hearts are heated for comparatively longer, to attain sufficient tenderness. The latter results in lower vitamin retention as indicated in Table 49.

As mentioned braising or simmering, boiling, roasting, etc., result in a sizeable loss of thiamin and riboflavin from the meat to the water. The retention in the meat may be as low as 30 to 40%. If the vitamin in the water is included in cooking, the retention is 46 to 90%. Riboflavin retention is generally higher than thiamin retention.

The results in Tables 48 and 49 were obtained in cooking unfrozen meat. The results from some experiments where frozen meat was used are given in Table 51. The data in parentheses are retention in relation to a comparable preparation from unfrozen meat. Lowest retentions are found for thiamin, 35% to 88%. Riboflavin retention was from 51% in pork loin to 134% in beef. Niacin retention varied from 50% in pork loin to 110% in beef liver.

Lee *et al.* (1950) and (1954) also analysed for pantothenic acid and pyridoxin. Only 59% retention of pyridoxin was found in pork loin where

Table 49. Vitamin retention in beef kidney and hog heart after preparation. After Noble (1970). Adapted from: Thiamine and riboflavin retention in cooked variety meats. Copyright The American Dietetic Association. Reproduced by permission from JOURNAL OF THE AMERICAN DIETETIC ASSOCIATION, Vol. 56: 225, 1970.

		% Retention			
		In meat		Total, incl. drip	
		Thiamin	Ribo-flavin	Thiamin	Ribo-flavin
Beef kidney	Braised	45	65	70	87
	Simmered	37	46	90	83
Hog heart	Braised	34	72	46	90
	Simmered	30	69	53	99

Table 50. Effect of freezing, thawing and cooking on vitamin content. After Lehrer *et al.* (1951).

Treatment	Retention in per cent of initial content		
	Thiamin	Riboflavin	Niacin
Frozen and thawed	85	109	81
Frozen, stored 3 months	78	71	92
6 months	78	71	92
Freshly cooked	44	66	83
Frozen, stored, cooked after thawing	36	52	59
Frozen, stored, cooked without thawing	51	63	52

otherwise comparatively high retentions were found. Thus, it is not unlikely that pyridoxin losses in preparing meats may be considerable.

Table 50 and 51 give retention in meat prepared from unfrozen raw material. In all cases lower vitamin content in the meat prepared after freezing than in the meat prepared without previous freezing was found. Only for thiamin and riboflavin does the difference appear to be significant, however.

Table 51. Retention of B-vitamin after freezing and cooking (In parentheses). Retention after cooking of the comparable, unfrozen product (untreated) given separately. Compiled by Lüthje (1982).

	Storage	Thiamin	Riboflavin	Niacin	Pantothenic acid	Pyridoxin	Preparation	Reference
Pig	6 months, −18°C	35 (79)	57 (86)	51 (61)			Roasting	Lehrer et.al., 1951
Long.dorsi	6 months, −26°C	37 (83)	51 (77)	50 (60)				
	Untreated	44	66	83				
Pig	6 months, −18°C	82 (93)	78 (83)	91 (93)	115 (129)	59 (86)	Roasting to temperature of 66°C	Lee et.al., 1954
Long.dorsi	Untreated	88	95	98	89	68		
Ham	Short time, −18°C	78	66				Boiling	Shestakov, 1976
Cured ham	2 months, −23°C	81					Roasting to centre temperature of 55°C	Bowers, 1979
Beef	10 months, −18°C	63 (99)	59 (61)	81 (87)			Roasting to centre temperature of 66°C	Lee et.al., 1950
Long.dorsi	Untreated	63	97	92				
	6 months, −18°C	59 (100)	84 (93)	85 (88)	78 (78)	88 (89)		
	Untreated	59	90	96	100	99		
Beef hind meat	a few days, −18°C	57	134	69	70		Roasting	Westerman, 1949
Beef liver	60 days, −18°C	51 (59)	70 (59)	110 (127)			Boiling	Kotschevar, 1955
	Untreated	85	94	86				

Fluctuating temperatures

Table 52 shows thiamin retention at constant or fluctuating temperatures, i.e. ±5°C, in ground beef patties, quoted by Kramer *et al.* (1976). As is seen, fluctuating temperatures, gave reduced thiamin retention, especially at the warmer temperatures. This may be compared to Gortner *et al.* (1948), who could not determine a reduced retention due to fluctuating temperatures after storage of roast pork when they investigated the effect of fluctuating storage temperatures on pork. *Longissimus dorsi* were stored at -18°, -12°C, and at temperatures fluctuating between -18°C and -12°C. In no case was a significant change found in the thiamin content. The authors did not find any difference in thiamin retention after 4 and 18 months at the storage conditions mentioned.

Table 52. Thiamin retention in frozen ground beef patty. After Kramer *et al.* (1976).

Storage	Thiamin retention	
	Constant temp.	Fluctuating temp.
3 months, −10°C	93	60
−20°C	97	70
−30°C	107	87
6 months, −10°C	63	60
−20°C	90	57
−30°C	90	87

Table 53. Thiamin retention after various treatments of ham loaf. After West *et al.* (1959)

	Thiamin retention
Raw	100
Prepared	72
Prepared, frozen 2 months, heated	73
Raw, frozen 2 months, prepared	73
Prepared, frozen 4 months, heated	72
Raw, frozen 4 months, heated	72

B-vitamin retention in frozen dishes

As indicated in Table 53, West *et al.* (1959) could not determine any loss in thiamin content during freezer storage of ham loaf. In general, losses seemed related to the preparation only, not to the freezer storage.

Rogowski (1976) determined the thiamin and riboflavin retention in four meat dishes after storage at -25°C for 6 months. The results are given in Table 54.

Table 54. Retention after 6 months' storage at -25°C. After Rogowski (1976).

	Retention	
	Thiamin	Riboflavin
Goulash	104	50
Marinated beef	107	103
Roast pork	77	116
"Bratwurst"	82	88

Zacharias and Bognár (1975) found that it would be most prudent not to store frozen ready to cook dishes longer than 3 months. Where they are stored longer, one should calculate with an estimated loss in nutritive value in any food intake calculation. These workers found thiamin and riboflavin retention at 73% and 83% respectively after 6 months' storage at -25°C in a series of dishes which are generally used in school meals in the Federal Republic of Germany.

Comparing cook-freeze and warmholding procedure

As mentioned above, there are practical advantages in using a cook-freeze system in preference to traditional warmholding in meal supply systems. The nutritional consequences of the two systems depend on the lengths of the warmholding period, and also on the type of oven which is used for heating the frozen dishes. Some results relating to this are given in Tables 55 and 56.

Table 55. Thiamin retention after cook-freeze-reheat and after warmholding of beef stew. After Kahn and Livingston (1970).

	Thiamin Retention
Prepared	100
Held warm for 1 hour	74
Held warm for 2 hours	68
Held warm for 3 hours	63
Frozen and thawed	96
Frozen, then heated in IR-oven	91
Frozen, then heated in microwave oven	95
Frozen, then heated in boiling water	85

Warmholding in this case means keeping the food at 82°C. The frozen dishes were heated to a temperature of about 90°C. The cook-freeze method gave for all three heating methods a better retention of thiamin than warmholding as determined by Kahn and Livingston (1970). The results from a similar experiment by Ang *et al.* (1975) are given in Table 56.

In this case, thiamin retention in warmholding of one dish, i.e. roast beef with gravy, was higher than what was found by Kahn and Livingston (1970). There was no significant loss after 30 and 90 minutes warmholding. Best retention for the frozen samples was obtained by heating in microwave ovens. The thiamin losses in warmholding of beans and Frankfurter sausages were significant but small for all three periods of warmholding, while freezing and heating in an infra-red oven did not result in a significant thiamin loss. Only when the cook-freeze system used heating in steam and warmholding for three hours, were significant riboflavin losses reported. Steam heating always gave the least satisfactory vitamin retention.

It is difficult to make general conclusions on the basis of the two studies, since retention seems to be dependent on the type of dish concerned. Nevertheless, the results do suggest that cook-freeze systems are preferable to warmholding if the latter is extended beyond 90 minutes.

Impact of meat freezing on the Danish diet

It was mentioned that meat makes significant and possibly at times decisive contributions to intakes of thiamin, riboflavin, pyridoxin and cobalamine. The data cited above in general show quite modest losses in these factors due to freezing, freezer storage or thawing of meat. In addition, some of these losses are partly offset by a correspondingly somewhat lower loss during

Table 56. Retention after cook-freeze-reheat (CF) and after warmholding (WH). After Ang *et al.* (1975).

	Roast beef with gravy		Frankfurter with beans	
	% retention			
	CF	WH	CF	WH
Partly prepared	100	100	100	100
Prepared	99	90	96	97
Held warm for ½ hour	93	89	93	98
Held warm for 1½ hours	92	91	86	93
Held warm for 3 hours	83	83	82	95
Frozen and thawed	94	94	97	95
Frozen and heated in convection oven	87	92	93	97
Frozen and heated in infrared oven held warm for ½ hour	87	93	94	97
Frozen and heated in microwave oven held warm for ½ hour	88	89	90	99
Frozen and heated in steam held warm for ½ hour	83	86	88	93

subsequent food preparation, cf. Table 50. It seems that, on the average, thiamin content may be reduced by as much as 50%, and riboflavin by as much as 30%, while the other B-vitamins are hardly affected. However, for pantothenic acid and pyridoxin, only few data exist.

It was suggested above, as a very rough estimate, that 32% of the meat in the Danish diet is exposed to one or several freezings and freezer storage periods. In very exceptional cases, freezing could cause losses of up to 50% of some of the vitamins compared to what would be obtained from unfrozen meat which had undergone a similar final preparation before serving. Extensive use of freezing could thus result in a not insignificant loss of some nutrients in the total diet, especially thiamin, about 18%. Considering the fairly adequate intake of these vitamins, which the Danish population has, see Table 29, this could hardly be a matter for serious concern. On the other hand, the impact of freezing is not insignificant, and therefore deserves attention in further studies of the freezing preservation of foods, especially if or where freezing meat is or gets to be the predominant method of meat preservation, possibly for special groups, e.g. elderly persons through cook-freeze systems.

More important, from a health standpoint, might be losses in pyridoxin, cf. Table 29, since intake of this factor from meat seems high and critical. Since some evidence has been found that pyridoxin is affected by freezing, this question merits close attention.

Another important finding is that there are losses, sometimes not insignificant, of nutrients during freezing, freezer storage, and thawing of meat. It seems that the public has a right to expect food preservation, food preparation and food distribution methods to be as gentle as possible with regard to the nutritive value of their foods. Therefore, there is ample reason to suggest that the effect of freezing on the nutritive value of meat be kept under constant review, especially until more is known and always where new processes or products are introduced, even though no public health concern appears to be involved.

Effect on food habits

In all deliberations regarding the effect of industrial processing on the nutritive value of foods, one must consider the effect which industrial processing has on food habits. Thus, freezing peas and other green vegetables may in many cases make these available at seasons or in areas where they otherwise would not be available. Thus, preservation by freezing may well facilitate access to a more balanced, more nutritious diet than would otherwise be available in many areas or in some seasons. This factor also is in need of further elucidation. Similarly, the fact that meat is frozen quite extensively may well mean that freezing is a means of making meat more readily available at lower cost and may thus contribute to increased meat intake.

However, there is little evidence that access to freezing and to frozen meat has affected meat intake in Denmark.

In general, it seems that problems of the nutritive aspects of frozen foods merit further study. This should concern itself not just with the fate of nutritive factors in the products, but also with the effect on general nutrition, which the availability of food freezing may have where it is done at the manufacturing or wholesale level in the form of freezing raw material for further processing, in catering, in the retail trade and in home freezing.

The PPP-Factors

As was indicated above, the author feels that in experiments with and considerations of frozen foods, exaggerated attention has often been paid to the effect on quality of freezing rate, storage time, and storage temperature and that more tests should be carried out on the influence of such factors as the nature and composition of the product itself, the process it undergoes, and the type of packaging in which it is frozen, stored and sold. The author first called attention to this in 1963 at the IIR Congress in Münich as published in Dalhoff and Jul (1965). The idea originated from the fact that in the series of experiments carried out in Albany, one case was recorded where fried chicken parts packed in packaging material of average quality resulted in a stability time of about 300 days at -20°C while approximately the same stability time was obtained for the same product in a package of high quality material at -8°C, see Fig. 47, cf. also Figs 29 and 30. Thus, it seemed that in such a case one might do as much or more for the quality of a frozen food by improving packaging as by major manipulation of temperature. To the author, this appeared all the more important because manufacturers may control product, product composition, processing, and packaging, but have little or no control over temperatures, at least not at the wholesale and retail level and in consumers' home freezers.

The author suggested that one should pay as much attention to three specific factors influencing keeping quality. They were described as product, process, and packaging, and Dalhoff in Dalhoff and Jul (1965) suggested that one operate not only with a TTT (time-temperature-tolerance) concept but, as mentioned above, also with a PPP (*product, process,* and *packaging*) concept.

Product

The nature and condition of the raw material used in freezing can have very

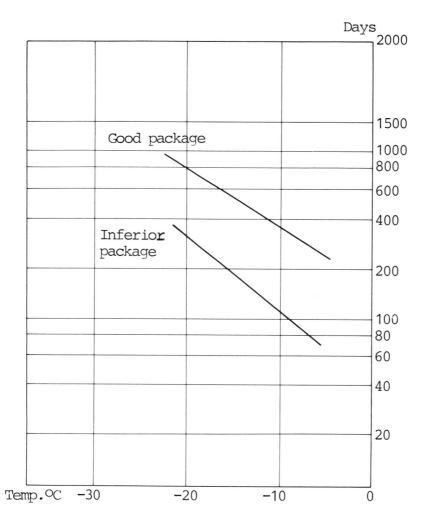

Fig. 47. Stability times for broiler parts in a very good and a poor package, both in commercial use at that time. After Klose *et al.* (1959).

considerable influence on the stability of the frozen product. Thus, Table 57 shows the difference in shelf life during freezer storage between two different qualities of pork, both commercially available. One is the result of a feed management based on avoidance of unsaturated fats in the feed and resulting in a rather low iodine number of the fat. Another is the result of corn feeding pigs resulting in a somewhat higher iodine number but also fully acceptable pork, actually preferred by many over the other product in the fresh state.

Table 57. Acceptability time for ground pork packaged in waxed paper, according to iodine number of fat. After Palmer *et al.* (1953).

Iodine No.	Acceptability time at –18°C in days
68.44	240
58.75	400

Table 58. Observed maximum storage life in months for plastic packaged chicken broilers in Denmark and recommended maximum storage in the Federal Republic of Germany. After Boegh-Soerensen (1975) and Ristic (1975).

	Denmark	FRG
–24°C	>30	7-8
–18°C	28	5-6
–12°C	15	3-4

However, the latter product is more susceptible to rancidity during freezer storage. Since rancidity is one of the limiting factors in the keeping quality of frozen pork - as in many other frozen products - the latter product has about half the stability and acceptability times as those of the former. Thus, such factors may be very important when freezing pork is considered.

Almost the same results as shown in Table 57 were obtained by Wismer-Pedersen and Anna Sivesgaard (1957), see Fig. 48.

Zaehringer *et al.* (1959) found that adding tocopherol to the pig ration considerably improved the storage life of frozen pork as measured both organoleptically and by TBA value and peroxide number.

Similarly, feeding chicken and turkey 0.1% tocopherol increased the shelf life of the frozen product considerably, cf. Mecchi *et al.* (1956).

For cod, MacCallum and Chalker (1967) found some cases where the condition of the fish at the time of capture influenced the shelf life of the frozen products, especially when refreezing was employed.

Carlson (1969) found that keeping fish deepchilled (superchilled) prior to freezing resulted in seriously reduced shelf life.

Winger (1982) found that electrical stimulation of lamb prolonged the shelf life of the frozen carcasses by about 30%. He found indications that variation between animals also might account for considerable differences in storage life of frozen lamb.

Findings such as these suggest that in comparisons between different experiments one should use great caution when comparing keeping qualities.

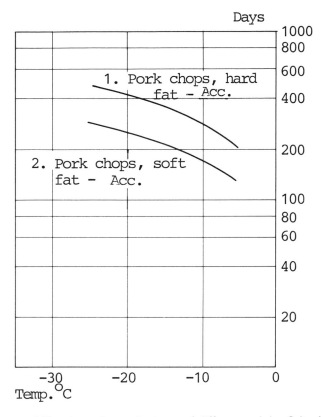

Fig. 48. Acceptability times for pork chops of different origin. Criteria: Accept-ability time for a decrease to a score of 6 in a 0-10 organoleptic scale (0 = very poor, 10 = excellent). Product: 1) pork chops from pigs fed barley and skimmed milk; 2) pork chops from pigs fed oily fish meal and garbage. Packaging: Aluminium foil. After Wismer-Pedersen and Anna Sivesgaard (1957).

In each case one needs an exact description of the type of products used. It is likely that some intercountry comparisons are indicated. Thus, Table 58 shows the acceptability time for frozen broilers determined in Denmark at the Danish Meat Product Laboratory (Boegh-Soerensen, 1975) compared with the acceptability time recommended for the comparable German product by the Federal Meat Research Institute in Kulmbach, Germany (Ristic, 1975). The difference is striking. It seems evident that the score used as a limit for acceptability in Germany was somewhat higher than that used in Denmark. Otherwise it is not likely that the Danish taste panel was much more tolerant than the German one. Animal husbandry specialists suggest that one explanation may be that there is a difference in feeding practices between

Table 59. The effect of formulation on percentage cooking loss from patties cooked nonfrozen, frozen and after thawing. Weight loss calculated as percentage of pre-cooked patty weight. After Nusbaum (1979).

	0% Frozen trim	50% Frozen trim	100% Frozen trim	Pre-rigor trim	Post-rigor trim
Cooked from unfrozen state	25.24	21.73	17.29	15.90	21.64
Cooked from frozen state	22.29	23.01	17.48	16.10	23.00
Cooked after thawing to 5°C	18.99	17.75	15.23	13.42	19.49
Overall treatment means	22.17	20.83	16.66	15.14	21.44

Denmark and the Federal Republic of Germany, and thus that the differences in keeping times observed may actually be due to such differences. In this case, it would be very useful to present samples of Danish and German origin and known histories to both the Danish and the German taste panels. It is to be hoped that such comparisons can be carried out in the future.

As is noted above, one has to accept, of course, that different panels have different acceptability concepts. All that is suggested is that one should have a clear indication as to the existence of such differences in order that, for instance, differences in raw material not be disguised by differences in taste panels and vice versa.

Nusbaum (1979) found an interesting product factor in experiments with ground beef patties. He obtained a much lower cooking loss if frozen patties were prepared from pre-rigor meat, cf. Table 59.

Ristić (1982) found that water chilled chicken broilers consistently were judged better than air chilled broilers when frozen.

An interesting example of the influence of the nature of the product on keeping quality was described by Steinbuch, Spiess and Grünewald (1977). They compared taste tests on frozen beans as they were carried out at a German and a Dutch Institute and found that any difference in taste testing technique was overshadowed by very significantly different keeping qualities due to differences in the products. In the Dutch experiment, beans of the "Record" variety were used; the German experiment used the "Jutta" variety. It seemed that when it came to shelf life of the frozen beans, that of the "Record" variety was about twice as long as that of the "Jutta" variety,

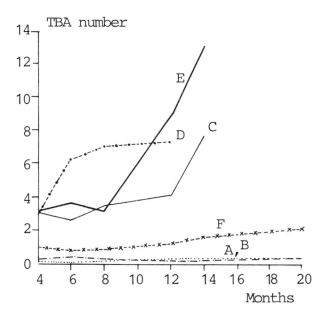

Fig. 49. Development of rancidity in frozen, cured, cooked pork. A, Nitrite + tripolyphosphate + ascorbate; B, Sodium chloride + tripolyphosphate + ascorbate; C, Nitrite + sodium chloride + ascorbate; D. Nitrite + sodium chloride + tripolyphosphate; E, Nitrite + sodium chloride + tripolyphosphate + ascorbate; F, Nitrite + sodium chloride + tripolyphosphate + double the permitted ascorbate. After Zipser and Watts (1967).

although conditions, e.g. blanching and packaging, were kept as identical as possible.

Marilynn W. Zipser and Betty Watts (1962) demonstrated a quite dramatic effect of product composition. Some of their results are given in Fig. 49. Curve A and B show the protective effect of ascorbate in combination with tripolyphosphate. Curve D shows that if sodium chloride is added, rancidity sets in very rapidly. If ascorbate is added in amounts of 550 mg per kg, this effect is counteracted as long as some ascorbate is left in the product, curve C. After that rancidity sets in, even more rapidly than in the product without ascorbate. However, when as much as 1.1 g per kg ascorbate is added, protection against rancidity seems to last throughout any normally occurring storage time, curve F. This is illustrated by comparison with Fig. 50. The effect of salt in a flesh product is seen also in Table 60 and Fig. 51.

It is important to underline that the very undesirable effect of salt is noticed mainly in the presence of air and absence of antioxydants. For vacuum-packaged smoked sliced bacon Lindeloev (1978) found mainly neutral stability, but a shelf life of 800 days in the temperature range of -8 to -60°C, cf.

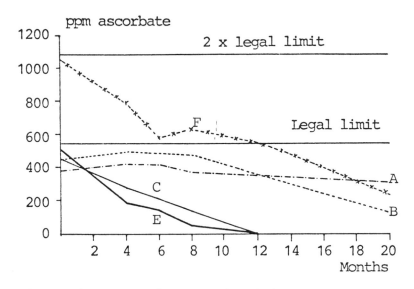

Fig. 50. Loss of ascorbate in frozen, cured, cooked pork. For explanations, see Fig. 49. After Zipser and Watts (1967).

Fig. 39. Vacuum-packaged sliced bacon is often sold in Denmark, and salt-cured herring, covered with brine and packaged in vacuum-bags and also frozen, is frequently sold in Sweden. Only the exclusion of oxygen from the package makes these practices possible.

On the other hand, one should be aware of the drastic changes in acceptability time which the addition of salt can cause. Thus, Table 60 suggests an acceptability time at -18°C for fresh, cooked pork in oxygen permeable films of 30 days, but only a few days if salt is added.

One should note that such effects may not be found by accelerated tests, which would often show longer stability times at say -5°C than at the projected storage temperature, e.g. -20°C, cf. Figs. 35 and 51.

The author knows of one case where the very prooxidative effect of salt was insufficiently realized. A manufacturer had for a considerable period produced a pork product, breaded pork chops. The products were individually packed in polyethylene film and packed in cartons for catering purposes. At one time, the customer asked that some salt be added to the batter. The appropriate modification of the batter was made and tested with satisfactory results. However, after some months, everything that had been manufactured according to the new specifications turned out to be rancid, and a very substantial economic loss was incurred. As mentioned, the product was packed in an oxygen permeable material. Thus, knowledge of such results as those obtained by Zipser, Kwon, and Watts (1964), Table 60, or Lindeloev

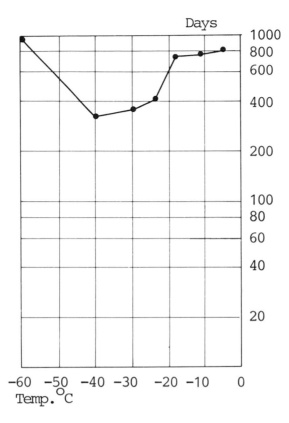

Fig. 51. Shelf life of vacuum-packaged, smoked Danish pork sausage (with binder. The salt content was not given; for that type of product, it is generally about 2%). After Lindeloev (1978).

Table 60. Effects of curing ingredients and antioxidants on cured and uncured pork stored at 0°F (-18°C). Criteria: Score of 4 in a 1-6 sensory score scale for odour. Product: Fresh ham, trimmed for fat and finely ground and cooked. Packaging: Polyethylene bags. After Zipser, Kwon and Watts (1964).

Product	Shelf life, days
Uncured	30
Uncured, 0.1% ascorbate added	360
Cured (2% NaCl, 0.03% NaNO2)	7-50
2% NaCl added	Very short
0.03% NaNO2 added	360
4% NaCl added	Very short
5.1% KCl	30
Cured, 0.01 BHA added	370
Cured, 0.1% ascorbate added	200-360

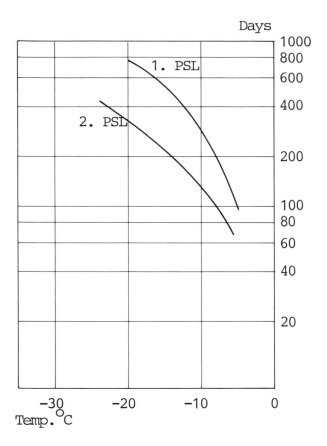

Fig. 52. Acceptability times for calves liver for a decrease to a score of -2 in a -5 to +5 organoleptic scale. Product: 1) calves livers of high degree of freshness, bacterial count of 30 000/g:2) calves liver after storage in the fresh state, bacterial count of 630 000/g. Packaging: coated, sealed carton; individual pieces wrapped in wax paper. After Dalhoff and Jul (1965).

(1978), cf. Fig. 35, would have made it possible to anticipate such an effect of a seemingly innocent product modification.

As mentioned above, the advantage of the Ottesen freezing method is that surprisingly little salt is absorbed in the tissue, e.g. of unpackaged fish. Yet, some salt uptake does occur and may lead to accelerated rancidity, cf. Fig. 60. This is probably why the method eventually was given up.

Many other product characteristics may, of course, be involved. A few will be mentioned here. Figure 52 indicates that a liver product of a high bacteriological standard is much more stable at freezer temperature than a

Table 61. Cooler storage of pork prior to freezing decreases shelf life of pork. (It may improve taste after shorter storage periods because of its ageing effect). After Harrison *et al.* (1956).

	Holding		Acceptability time
7 days	at 40°F	(4.4°C)	ca.120 days
7 days	at 30°F	(-1.1°C)	ca.133 days
3 days	at 30°F	(-1.1°C)	ca.156 days
1 day	at 30°C	(-1.1°C)	ca.212 days

liver product of lower bacteriological quality. The differences in bacterial count were simply due to one product being stored somewhat longer prior to freezing than the other. Both products were fully acceptable organoleptically. It is likely that the difference is due to the activity of bacterial enzymes present in the product which had been stored fresh for a brief period before freezing. As is known, enzymatic activity, cf. Table 20 and Fig. 22, is reduced but by no means blocked at conventional freezer storage temperatures. Table 61 gives a similar example for pork.

This last example leads to some interesting observations with regard to the freezing of products which normally need to be aged, e.g. better qualities of beef. One cannot assume that consumers will age a product after thawing, prior to consumption. In fact, such ageing would not be as effective as ageing the product prior to freezing. At any rate, ageing of beef prior to freezing is a technique of long standing commercial practice. However, such ageing results in the formation of some bacterial enzymes, which penetrate into such parts of the beef which may not be trimmed off prior to freezing. Therefore, an aged beef product normally shows a lower stability during freezer storage than an unaged product. Obviously, in this case there will be a trade-off between the advantages of ageing against the disadvantages of the reduced keeping time. The subject needs to be further elucidated by experimentation, cf. Fig. 53. In one experiment Winger and Pope (1981) demonstrated that conditioning of lamb carcasses, i.e. a holding period of 24 hours at +6°C before freezing resulted in the onset of rancidity very considerably earlier than in non-conditioned lamb. Thus, while conditioning is carried out because it improves tenderness, one has to consider also other quality factors, e.g. rancidity as in this case, where an adverse effect is noted. When electric stimulation, accelerated conditioning and slow freezing is used, as is now general practice in New Zealand, this adverse effect on rancidity may be avoided.

Estimating the keeping time for fish is particularly difficult because differences exist between fish of the same species from different areas. Besides, fish is rarely frozen immediately after capture. Storage times and conditions prior to freezing may have a marked influence as indicated for haddock by Dyer and Peters (1969) in Fig. 53.

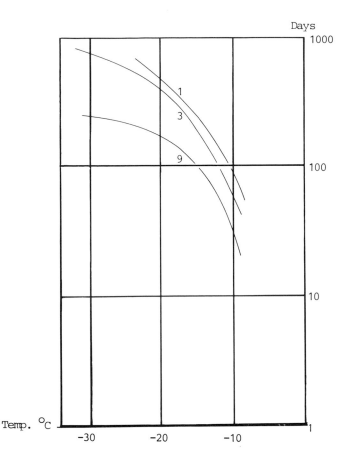

Fig. 53. Shelf life of haddock iced for one, three or nine days prior to filletting, packaging in 6.1 kg blocks, and packaged in waxed cartons and frozen. After Dyer and Peters (1969).

Similar results were reported for coalfish by Bramsnaes (1969). The matter is further complicated by the fact that the method of packaging may have a very considerable influence on shelf life, cf. Table 64 and Fig. 60. Factors influencing the shelf life of frozen fish are further discussed by Connell (1969) and by Connell and Howgate (1968, 1969 and 1970).

The use of additives to gravies and puddings had some effects in experiments reported by Hanson *et al.* (1957 and 1958). Another example of the influence of product and product composition is seen in Fig. 54.

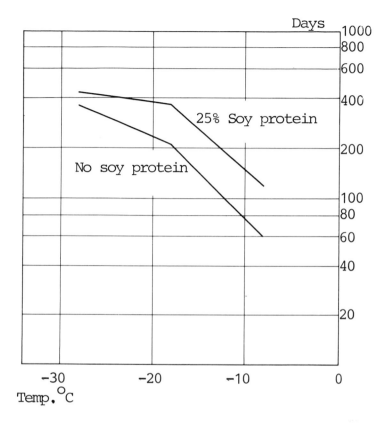

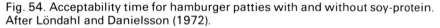

Fig. 54. Acceptability time for hamburger patties with and without soy-protein. After Löndahl and Danielsson (1972).

Guadagni (1969) notes that adding sugar to some fruits before freezing may improve their shelf life but mainly at relatively warm freezer storage temperatures.

Process

Frozen products have often undergone various degrees of or types of processing prior to freezing. For instance, one often has a choice as to whether a product should be cooked prior to freezing or whether cooking should be carried out by the user.

One obvious example is that of blanching vegetables. Here, normal practice for most vegetables is to blanch the product, i.e. actually precooking the product prior to freezing. It is found that the destruction of enzymes, which is

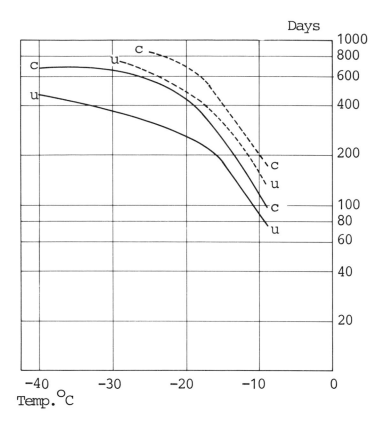

Fig. 55. Stability (- - -) and acceptability (- - -) times for uncooked (u) and cooked (c) Danish pork sausage (with binder), ("Medisterpølse"), packed in freezer cartons lined with coated heat-sealed paper liner. After Boegh-Soerensen (1967).

a result of blanching, increases the keeping quality during freezer storage very substantially.

Blanching is in itself a very important process factor and many differences in shelf life of frozen vegetables and some fruits are due to differences in the blanching procedures used.

Lately, this practice has been somewhat questioned. It seems that better organoleptic qualities may be obtained by improved packaging, e.g. vacuum packaging, and avoidance of blanching, provided better packaging and shorter keeping times can be accepted. This subject is in need of much further investigation.

Philippon (1981) points out that it is often necessary, e.g. for reasons of process capacity, to freeze fruits and vegetables in bulk at harvest time and to

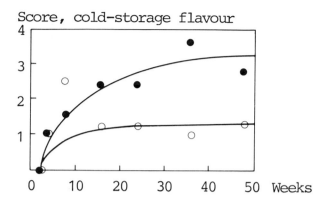

Fig. 56. Development of cold-storage flavour during storage at -13°C (fresh fish). -o-o-, cooked; -●-●-, uncooked. Score: 0 = no storage flavour; 4 = strong storage flavour. After Tatterson and Windsor (1972, reproduced by courtesy of MAFF-Torry Research Station).

leave peeling, further treatment, even blanching, to some convenient time, often several months later. Even where the process itself is finished immediately after harvesting, e.g. when peeled and blanched peas are fluid bed frozen, the product may be bulk stored for packaging in retail packages or for use in some precooked dish many months later. Little is known about the quality aspects of such procedures.

Generally, fruits and vegetables have a significantly longer shelf life if freezer storage temperatures are decreased. This is of special significance in cases where vegetables are frozen without blanching for further processing later. Thus, unblanched carrots may keep for 3 months at -20°C but for 12 months at -45°C. One may therefore avoid changes during freezer storage for unblanched products by adopting a very cold storage temperature, e.g. -30 to -35°C. This would be possible since such products would not enter into the regular freezer chain but could be stored in freezer storage rooms set aside for this specific purpose and would not need to be transported or transferred until their use in further production.

This question is discussed in considerable detail in the proceedings from a meeting on that subject in Paris, 19 November, 1981, cf. Philippon, Ulrich and Zeuthen (1982).

Figure 55 shows that similar considerations regarding heat treatments are indicated for other products, thus, in this case the cooked sausage kept considerably better than the uncooked one. Similarly, Fig. 56 shows a case where an unpleasant "cold storage" flavour developed more slowly in a product which had been cooked prior to freezing than in the uncooked product. A similar effect was reported by Dawson (1969) for fried chicken.

Table 62. Stability times in days at triangle tests for boysen-berries. After Guadagni, quoted by Philippon (1981).

	−18°C	−12°C	−7°C
Without sugar	405	125	45
With sugar	650	160	35

Fruits are often better preserved when frozen in syrups. Table 62 shows that this is particularly true at cold temperatures.

These results were obtained by Guadagni using triangle tests comparing with the control kept at -29°C. This, however, may also be one of the cases where the testing method could lead to erroneous conclusions. The products kept at -18°C may have undergone changes parallel to the controls, while those maintained at -7°C may have changed in another, but not necessarily more undesirable, direction. This effect might well be most pronounced for those products which were frozen in sugar and where, especially at warm temperatures, not much water was frozen out as ice in the tissue, cf. Table 10. This is another area in need of further investigation.

Philippon and Rouet-Mayer (1978) and Philippon *et al.* (1981) point out that peaches are best suited for further processing if unblanched, stored colder than -20°C and peeled before thawing, at about -4°C. Kozlowski (1977) also found cases where blanching of vegetables before freezing could be eliminated. As mentioned by Strachan (1983), blanching is unnecessary and undesirable in the case of onions.

In this context, various treatments may also be considered. Thus, Fig. 57 indicates that treating a fatty fish product with an antioxidant prior to freezing will increase its keeping quality very substantially. The same is the case for frozen, cured meat products as mentioned above, cf. Table 60.

Winger and Pope (1981) found that mincing lamb prior to freezing seemed to result in considerably reduced shelf life.

While the immediate effect of the freezing process as such on product quality is generally rather insignificant, it is worth noting that it may have some significance for the shelf life of the frozen product. Thus, Lindeloev (1978), as shown in Fig. 12, demonstrated that freezing to an end point of -40°C might for sliced, non-vacuum-packaged bacon result in a shelf life of only 37 days at -15°C, while freezing it to -15°C only resulted in a shelf life of about 80 days when the product was stored at that temperature.

One aspect, which so far has received little attention, may be classified under this heading, namely the fact that most laboratory experiments are carried out with very fresh raw material, processed, packaged and frozen immediately. In actual practice, this is often not even the predominant method of product preparation. Thus, fish is often frozen in blocks at sea, freezer

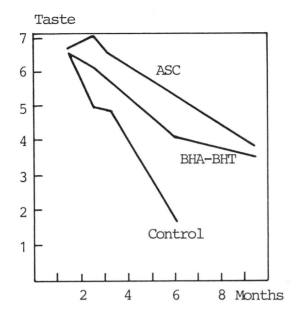

Fig. 57. Taste tests of trout at -10°C. Controls and samples treated with ascorbic acid (ASC) or BHA-BHT. After Aagaard (1968).

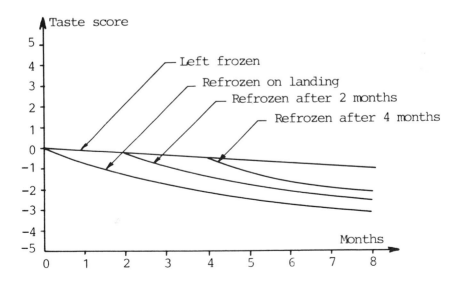

Fig. 58 Symbolized quality loss in thawing, filleting and refreezing cod. (Andersen, Jul and Riemann, 1966).

stored aboard and ashore and then tempered (partially thawed), cut up into fillets or used for frozen fish sticks. Dyer *et al.* (1962) made some experiments regarding this, cf. Fig. 58. In later experiments with both cod and redfish, MacCallum and Chalker (1967) found little effect of refreezing on shelf life.

As mentioned above, many frozen meat and poultry products are prepared from raw materials which have been frozen. This may affect storage life somewhat.

Packaging

As already touched upon, the PPP factor which often is of the greatest importance, is packaging. Figure 47 from the "Albany" experiments indicated a striking effect of a good versus an inferior packaging. It is interesting that already Plank, Ehrenbaum and Reuter (1916) mentioned "anaerobic" packaging as the second most important factor for the quality of frozen foods.

In 1959, Bramsnaes and Soerensen (1960) reported the data given in Table 63, showing the much improved keeping quality of fatty fish when it is vacuum-packaged in oxygen impermeable film.

Löndahl and Danielsson (1972) found the same for vacuum-packaged shrimp as seen in Table 64.

Table 63. Flavour scores for frozen trout at various storage periods and two types of packaging, in months at -20°C. Flavour score: 10 to 0 with below 4 being considered unfit for sale. After Bramsnaes and Soerensen (1960).

Time	Vacuum packaging	Oxygen permeable packaging
1	7.8	6.5
2	7.4	6.6
4	6.5	4.5
6	7.1	3.8
9	5.5	1.8

Table 64. Acceptability times in months at -18°C, according to taste of individually frozen shrimps. After Löndahl and Danielsson (1972).

8% glaze, vacuum packaged	10
4% glaze, vacuum packaged	8-9
Vacuum packaged	6-7
Packaged in polyethylene film	3-4

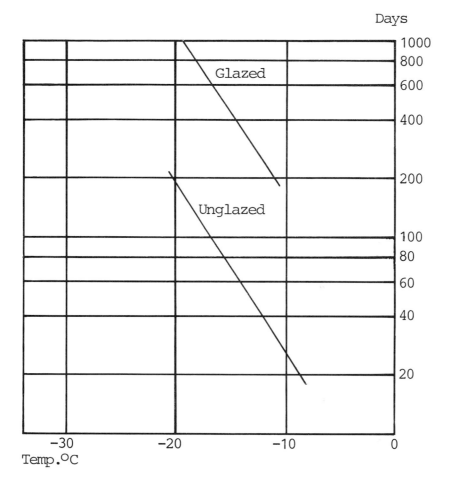

Fig. 59. Acceptability times for bulk packaged frozen hamburgers. After Löndahl and Aaström (1972).

Löndahl and Aaström (1972) indicated that a similar effect is obtained by such a simple packaging method as glazing. Thus, glazing hamburgers increased shelf life by 100-400%. Their results are summarized in Fig. 59. Fig. 60 shows a similar effect of glazing fish. Banks (1935) made findings very similar to those shown in Fig. 60.

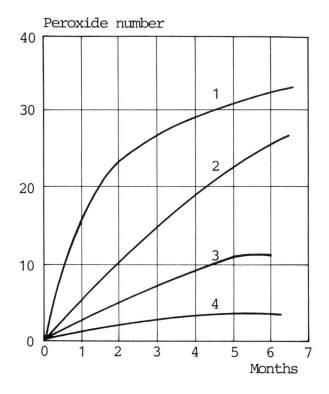

Fig. 60. Rancidity, measured by peroxide number, in herring. 1, Brine frozen, unglazed; 2, Brine frozen, glazed; 3, Air blast frozen, unglazed; 4, Air blast frozen, glazed. After Aagaard (1968).

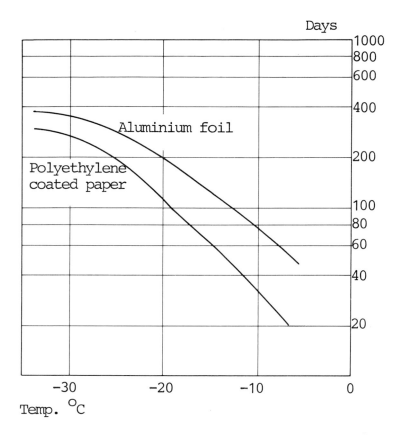

Fig. 61. The effect of various packaging materials on shelf life of individually packaged hamburger meat. After Löndahl and Danielsson (1972).

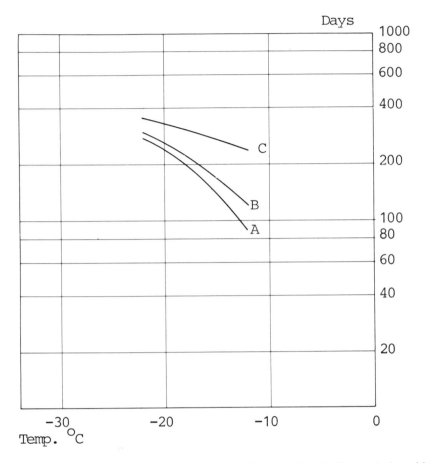

Fig. 62. Acceptability time for pork chops. A: Wrapped in PE film and placed in master cartons, B: same as A, but a PA/PE laminate film, and C: vacuum packaged in PA/PE laminate (same material as B). After Boegh-Soerensen (1982a).

For individually packaged hamburger meat, Löndahl and Danielsson (1972) found the data reproduced in Fig. 61. For pork chops, Boegh-Soerensen (1982a) found the data shown in Fig. 62. In this, as in Figs 30 and 61, it will be noted that packaging appears to be more important at warm storage temperatures, e.g. may be beneficial mainly during a product's time in a freezer retail cabinet.

Winger and Pope (1981) found improvement in the storage life of frozen lamb when the product was anaerobically packaged. The effect was not very considerable, however, probably because the shape of the lamb carcasses makes complete exclusion of air impossible.

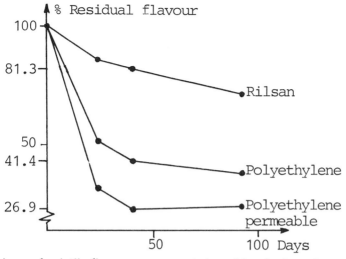

Fig. 63. Loss of volatile flavour components in unblanched parsley, at -16°C. After Philippon (1981). (Rilsan is assumed to be a polyamide resin).

Above reference is made to the very unsatisfactory keeping quality of frozen cured pork, when it is packaged in permeable material, cf. Table 25 and Fig.35. It is for this reason that such products are always vacuum packaged when they are sold frozen, cf. Fig. 39. In this technology, however, it is very important that no leakers develop. Riordan (1976) has shown how rancidity, i.e. measured by peroxide value or TBA-value, very quickly sets in in leakers.

One important observation was made by Boegh-Soerensen (1975), testing whole broilers and broiler parts, cf. Fig. 29, where the quality of the package has a measurable although rather limited influence on the quality of whole broilers. This may be compared to Fig. 30, where Boegh-Soerensen (1975) found a very significant influence of a good package on broiler parts. The explanation is probably that a broiler has a rather awkward shape for packaging because of its large body cavity. Obviously, it cannot be vacuum packed and air cannot be excluded. Thus, packaging in water vapour and oxygen impermeable material is of limited value, and has little measurable effect, at least unless packaging in an inert atmosphere could be used. On the other hand, vacuum packaging may provide a rather tight fit for broiler parts, cf. Fig. 30.

The importance of the packaging factors and an illustration of what can be obtained by improved packaging is seen in the fact that most data regarding keeping times for frozen food are generally considered obsolete from a commercial point of view if they are more than two to three years old as stated by M. Sanderson-Walker, UK, in 1980. This is because of changes in films, packaging methods, etc.

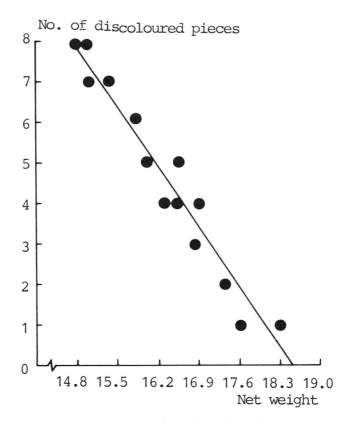

Fig. 64. Relation between net weight and number of discoloured pieces of peaches in cartons of the same volume at -6.7°C. After Guadagni, Nimmo and Janson (1957).

One fact, which has received insufficient attention, is that packaging materials should normally not only be impermeable to water-vapor and oxygen, but also to volatile flavour substances. Figure 63 shows an example of the influence of packaging material in this regard. In this respect, freezing loose products in a gravy or dressing is often an advantage, simply because the latter may protect against the evaporation of volatile components.

Package fill is another important component; sometimes, as in Fig. 64, because it reduces exposure to air. Boegh-Soerensen (1982a) obtained comparable results, e.g. filling a package up with sugar syrup, a gravy, or a dressing improves keeping time.

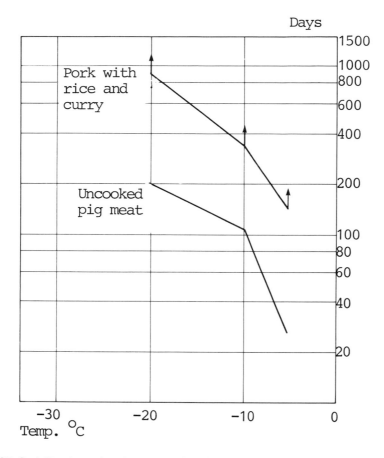

Fig. 65. Stability time of pork untreated, and in a ready to heat, vacuum-packaged dish. (Dalhoff and Jul, 1965). An arrow in the figure indicates that all samples were used before the stability time was reached.

Combined PPP-effects

Some results of possibly all three PPP-factors are given by Dalhoff and the author, obtained 1961 and published 1965, see Fig. 65.

In this case, keeping quality of a well prepared and well packaged dish containing pork was compared with the quality of uncooked pork for which few PPP-factors had been used for improving shelf life.

Similarly, Tables 21 and 22 summarized stability and acceptability times for a variety of products, determined by the same authors. These data show that by combining PPP-factors intelligently, one can obtain much longer keeping times than often believed possible. Thus, when these data were submitted as a paper for the 11th International Congress of Refrigeration in Munich in 1963, the paper was originally rejected because it was assumed that the results somehow were erroneous, some even suspected that results had been manipulated. The reason for this reaction was that it was then a generally accepted view that one could never obtain acceptability times for pork exceeding about 200 days at -20°C. Results postulating recorded acceptability times of 200 days and more for pork products at -10°C or warmer were therefore considered unacceptable. The explanation was simply that one had previously not examined the effect of an optimum combination of the three PPP-factors. In fact, the authors were very surprised themselves and ran out of samples at the various test temperatures because the products were found unaltered even at the longest storage times for which samples had been provided and where it was anticipated that the product would be past any conceivable keeping time. In fact, these were the experiments which led the author to believe that the frozen food industry, by carefully manipulating the PPP-factors, could obtain a much-improved shelf life of its products.

Food legislation

It seems that food control authorities also need to give as much recognition to the PPP-factors as to time and temperature. However, so far one of the few references found in legislation or regulations for frozen foods pertaining to PPP-factors seems to be that the raw material should be of good quality, i.e. presumably having a low microbiological count. However, this requirement is probably stipulated because of a general hygienic concern with regard to the finished product and not based on a desire to improve shelf life in the frozen state. Another reference is sometimes found in a requirement that the package for frozen foods should be air-tight and impenetrable by microorganisms. However, this is presumably also due to a traditional hygienic concern. One wishes that frozen foods should not be contaminated in storage, handling, and distribution. This concern is somewhat academic since, together with canned foods, frozen products are among those best protected against adverse bacterial contamination, in the case of frozen foods because of the inability of

Table 65. Temperatures required in France for various frozen products. After Delaunay and Rosset (1981).

Products labelled "surgelé"	–18°C
Fish and prepared dishes	–18°C
Poultry	–12°C
Meat designated "congelé"	–10°C
Thawed and refrozen products	–18°C

microorganisms to develop in the product in the frozen stage. There is a much greater need for a similar protection of unfrozen products, but similar demands are rarely made for those. As stated by McBride and Richardson (1979) and Ware (1973), most frozen food legislation appear unnecessarily preoccupied with temperature requirements. In this, it fails to give appropriate attention to product, process, and packaging factors. In actual fact many legislative and administrative requirements for food freezing and the frozen food trade seem to have limited relation to product quality, wholesomeness or safety, because of their failure to take these factors into account.

Table 65 shows that in France, some attempts are made towards a somewhat more differentiated approach, at least as regards temperature requirements.

Need for TTT-PPP data

In estimating the keeping potential of various products, it appears useful if more, and better defined, time-temperature data could be determined for each product to be considered and that these would properly take into account the PPP-factors.

In Figs 66 - 68 are reproduced the findings from various TTT-experiments as they were summarized by Bengtsson *et al.* (1972). Quite similar summaries were made by Spiess (1980) in the above mentioned unpublished report for the purpose of Codex Alimentarius work. A few of Spiess' many data sheets are shown as Figs 69 -75.

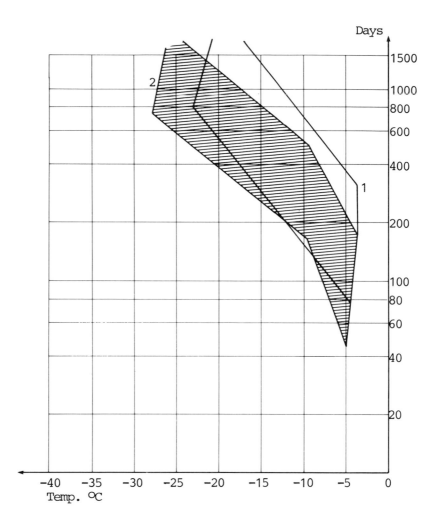

Fig. 66. Acceptability areas for lean meat and ready-to-heat dishes with lean meat in gravy (1) and fatty meat and ready-to-heat dishes with fatty meat in gravy (2). After Bengtsson *et al.* (1972).

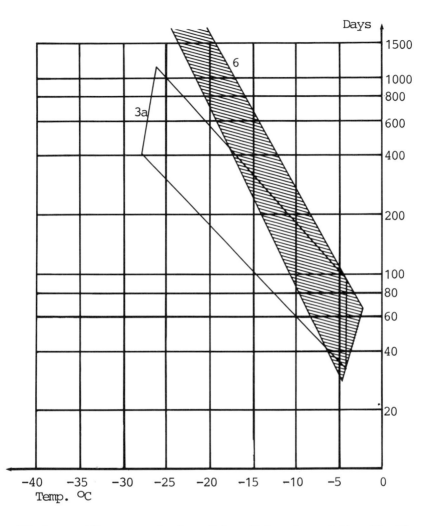

Fig. 67. Acceptability areas for lean fishes (3a) and vegetables (6). After Bengtsson *et al.* (1972).

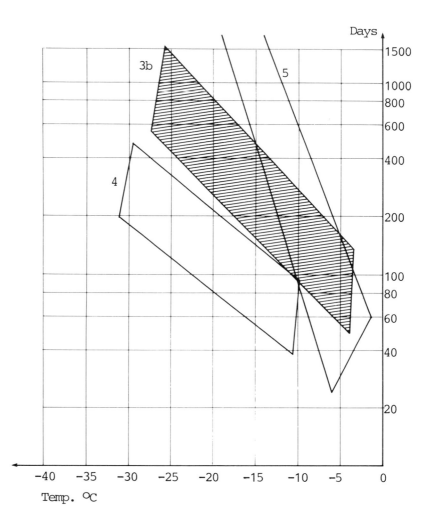

Fig. 68. Acceptability areas for ready-to-heat dishes without gravy (3b), fatty fishes (4), and fruits and berries (5). After Bengtsson *et al.* (1972).

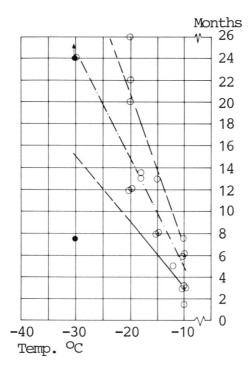

Fig. 69. Example of data sheets, stability time (HQL), for beef, collected by Spiess (1980).

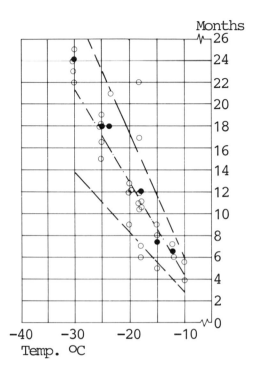

Fig. 70. Example of data sheets, shelf life or acceptability time (PSL), for beef, collected by Spiess (1980).

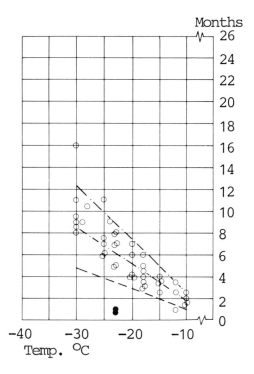

Fig. 71. Example of data sheets, stability times (HQL), for codfish, collected by Spiess (1980).

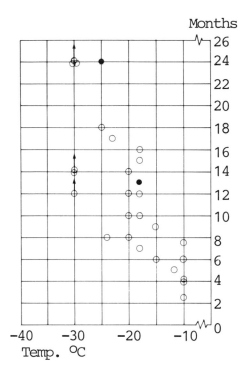

Fig. 72. Example of data sheets, acceptability time (PSL), for green beans, collected by Spiess (1980).

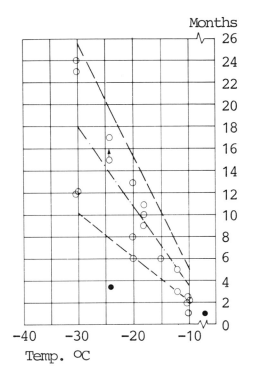

Fig. 73. Example of data sheets, stability times (HQL) for green beans, collected by Spiess (1980).

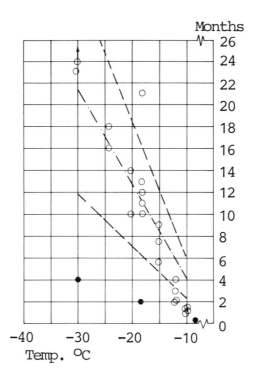

Fig. 74. Example of data sheets, shelf life or acceptability time (PSL), for strawberries, collected by Spiess (1980).

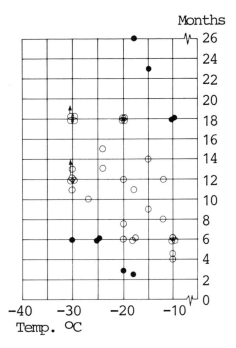

Fig. 75. Examples of data sheets, stability time (HQL), for ready-to-heat meat dishes, collected by Spiess (1980).

The enormous spread between findings at various laboratories and in various experiments as seen in these illustrations makes such data difficult to use. This divergence between data is partly due to different quality criteria and, and probably more frequently, to often unrecorded differences in PPP factors. It is obvious that the whole field is in need of further experimentation. In such, there is a need for very close attention to and description of all the PPP-factors, and also exact indication of the quality criteria used in shelf life tests. Reindl, Grossklauss and Busse (1983) suggest some guidelines for obtaining such data.

The urgency of such work needs to be considered with attention to the fact that many governments, Codex bodies, etc., are presently introducing or considering legislation or rules which may limit the permissible last day of sale after packaging (freezing) of various products in their respective countries or areas of responsibility. Also, the International Institute of Refrigeration (IIR) is periodically revising its recommendations for the handling and storage of frozen foods. In all of this work in which the author has also been involved, it is quite evident that insufficient experimental and empirical data exist to support recommended maximum acceptable storage times for frozen foods. Especially considering that the influence of product, process, and packaging factors can be enormous it seems unfortunate if references to some of those are not made. Thus, in the last available edition of the IIR recommendations, storage times are given without reference to any PPP-factor. Also, it seems evident from the above that the assumption of a constant Q_{10} at temperatures below say -5°C exists in many circles, while in reality few products adhere to this model. Thus, there is a great danger of placing undue restrictions on the sale of frozen products and frozen retail packaged foods without it leading to any quality benefit for the consumer but conceivably resulting in more expensive products.

Estimated Practical KeepingTimes

Additivity of effect of various episodes

Experiments carried out both at the USDA Western Regional Research Center and by the Danish Meat Products Laboratory confirm that normally when frozen foods are exposed to different temperatures, the cumulative effect on product quality is the same regardless of the sequence of temperature experiences, i.e. the effects are both additive and commutative. For instance, a product stored at -20°C for three months, and subsequently at -11°C for two months, will normally have the same quality as the same product which has experienced the same temperatures and storage times but in the reverse sequence. This obviously leads to a great simplification when it comes to estimating the effect on quality of various time-temperature experiences of frozen foods.

Justification for using this assumption as a working hypothesis is given by many workers.

Thus, Dietrich *et al.* (1959) studied the effect of varied sequences of storage at -17.8, -12.2, and -6.7°C in an experiment covering about 13 000 packages of frozen beans. The conclusion was that whether the warmer temperature came first, last or was interrupted by the colder temperature storage, the total deterioration was about the same when evaluated by flavour and colour judgements or measured by a certain loss in chlorophyll or ascorbic acid.

Similarly, Boggs *et al.* (1960) found for frozen peas the same flavour and colour changes measured subjectively and the same changes in chlorophyll, ascorbic acid and colour, measured objectively, after the following treatments:

a) 4 months at -12.2°C followed by 4 months at -17.8°C
b) 4 months at -17.8°C followed by 4 months at -12.2°C
c) 2 months at -12.2°C followed by 4 months at -17.8°C, and then another 2 months at -12.2°C.

They also found that the effect of regularly fluctuating temperatures was nearly the same as that which would result from storage for the same period at the equivalent steady temperature, the latter being calculated by Schwimmer and Ingraham's formula (1955). This was true for a 24-hour sine-wave cycle over the following temperature range: -20.6 to -15.0°C; -23.3 to -12.2°C; and -17.8 to -6.7°C. They also exposed frozen peas to temperature patterns simulating commercial practice in handling and distribution; from these data they calculated, by the method given below, the number of days at -17.8°C that would result in the same deterioration and found excellent agreement with the results of actual storage tests.

Guadagni *et al.* (1957b) found that for strawberries frozen with sugar, the following treatments gave identical results on flavour change and in ascorbic acid losses:

a)   6 days at -6.7°C followed by 3 months at -17.8°C
3 months at -17.8°C followed by 6 days at -6.7°C

b)   6 days at -6.7°C followed by 1 year at -17.8°C
1 year at -17.8°C followed by 6 days at -6.7°C

c)   12 days at -6.7°C followed by 1 year at -17.8°C
1 year at -17.8°C followed by 12 days at -6.7°C
6 days at -6.7°C followed by 1 year at -17.8°C, and then another 6 days at -6.7°C.

They also found that regularly fluctuating temperatures resulted in changes that were not significantly different from those that were found at the equivalent steady temperatures. The same was found for ready-to-cook cut-up chicken by Klose *et al.* (1959), except for frost formation, as mentioned below.

Hanson and Fletcher (1958) stored frozen turkey dinners and turkey pies at -6.7°C for ½, 1, and 1½ months followed by storage at -12.2°C for three months and compared them with samples exposed to the same conditions in reverse order. No flavour difference nor difference in peroxide value were found between corresponding samples.

Guadagni *et al.* (1957) showed that for peaches in sugar syrup, even when temperatures above 0°C were included in the temperature history of the product, the deteriorative effect of various temperature periods was cumulative. They found no specific effects from cyclic fluctuating temperature other than those that were to be expected from the total temperature history, calculated as indicated below.

McColloch *et al.* (1957) concluded after a very exhaustive investigation that the effects of time-temperature treatments on flavour in orange juice concentrates are strictly cumulative, even when temperatures above 0°C are included.

Table 66. Freezer storage management of cartons with hamburgers. Time indicated in weeks. a, b, and c indicates order of exposure to time and temperature episode. The samples in each group did not differ significantly from each other when tested in triangle taste tests. After Boegh-Soerensen and Hoejmark Jensen (1978).

	A	B	C	D	K	L	M	O	Q	R	S
–12°C	17b	17a	26a	26b	26c	26a	26b	21a	21c	21b	21c
–18°C								12b	12a	12a	12b
–24°C	26a	26b	26b	26a	42b	42b	42a				
–30°C								17c	17b	17c	17a
Freezer dis-play cabinet					4a	4c	4c				

Many of these earlier findings were reviewed by van Arsdel (1969).

Boegh-Soerensen and Hoejmark Jensen (1978) tested hamburgers packed in polyethylene bags, each containing four hamburgers, six bags placed in an outer carton. They were freezer stored according to the temperature management system indicated in Table 66. Since the samples in each group did not differ significantly when taste tested in triangle tests, it is concluded that in these cases, shelf life loss was practically the same regardless of the order, in which the samples were exposed to the various time-temperature episodes. When the stability and acceptability curves given in Fig. 76 were used, the actual loss in shelf life corresponded well to that at which one would arrive using the method of calculation indicated in Table 67.

Exceptions to the rule of additivity

There are some cases where the total effect of various temperature experiences may not be independent of the order in which they occur or of the nature of the temperature history.

Thus, very widely fluctuating temperatures may cause freezer burn or in-package desiccation. Klose *et al.* (1955) found that for turkeys in polyethylene bags the frost formed in the bag was 0.33% for one year's storage at temperatures fluctuating from -23.3 to -12.2°C over a 24-hour period. At constant temperatures, the corresponding losses were as follows: At -23.3°C about 0.07%; at -17.8°C about 0.17%; at -12.2°C about 0.20%.

Several foods with a high content of solubles, i.e. fruits frozen with sugar, ice cream, etc., have a low freezing point. It is not clear if passage or at least repeated passage of the freezing point may not cause some change that is not strictly cumulative. Even in some of these cases, however, the assumption of

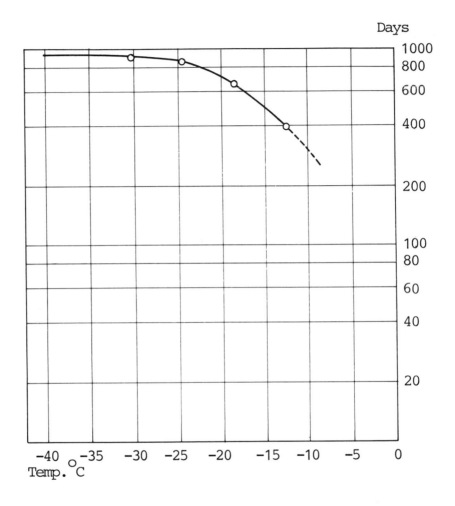

Fig. 76. Acceptability times of frozen pure beef hamburgers, packed 4 each in polyethylene bags, which were placed 6 each in outer cartons. After Boegh-Soerensen and Hoejmark Jensen (1978).

the effect being completely additive is justified as indicated above.

Where the colloidal nature of a product is affected, the effect of time-temperature treatment may not be additive. Hanson *et al.* (1957) found that for the texture of starch-thickened white sauce and egg-thickened pudding a fluctuating temperature, -23.3 to -12.2°C, was considerably more damaging than storage at the equivalent steady temperatures, calculated according to the method suggested by van Arsdel and Guadagni (1959).

When growth of microorganisms occur, the effect of time-temperature

Table 67. Calculation of shelf life loss for frozen hamburgers. Dwell times and temperatures in the freezer chain, estimated from Tables 73, 74 and 85 and Figs 80, 84 and 96, and acceptability time from Fig. 76.

Producer's freezer storage	100 days	at	−24°C	: 100	÷ 860	= 0.116
Transport	2 days	at	−12°C	: 2	÷ 390	= 0.005
Wholesaler's cold store	40 days	at	−24°C	: 40	÷ 860	= 0.047
Transport	1 day	at	−12°C	: 1	÷ 390	= 0.003
Retail cabinet	9 days	at	−17°C	: 9	÷ 500	= 0.018
Transport	30 min.	at	−5°C	: 0.02	÷ 150	= 0.000
Home freezer	12 days	at	−17°C	: 12	÷ 390	= 0.031
Total						0.220

treatments is unlikely to be additive. Michener *et al.* (1960) showed that bacterial growth normally takes place above -4°C in frozen peas, beans, cauliflower, and spinach. Therefore, deterioration during low storage temperature is likely to be faster if preceded by storage above -4°C, and such an experience prior to storage at colder temperatures will be more detrimental than when it takes place at the end of the storage periods.

The author (1960) discussed various factors which may interfere with additivity and concluded that in most cases occurring in practice, the effects will be additive.

Calculating shelf life losses

In general the author's experience confirms the view that, for the time-temperature experiences which frozen foods normally go through in the normal freezer chain, applying the rule about the additive and cumutative effect of various storage episodes will lead to reasonably accurate estimates of shelf life loss. Table 67 gives a calculation of the cumulative effect for frozen hamburgers of a passage through a frozen food distribution system. The calculation is based on Fig. 76 and is carried out as follows:

The product was first stored at a temperature of -24°C. According to Fig. 76, at this temperature the product has a shelf life or acceptability time of 860 days. This means that one day at this temperature would result in 1/860 of that degree of shelf life loss which makes the product no longer fully acceptable. The effects of transport and retail cabinet storage are calculated in the same way, in both cases using the maximum temperature not exceeded by more than 5% of the products in that episode of the freezer chain as product temperature for the purpose of the calculation. The full calculation is shown in Table 67.

It will be seen that the time-temperature experience has led to a product which is changed, but which is still well acceptable, i.e. acceptability time was not reached. This method of calculation, originally proposed by Hicks (1944),

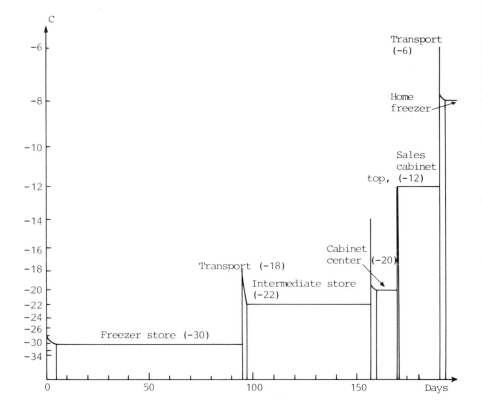

Fig. 77. Fate of frozen cod fillets package during distribution. Area under curve corresponds to shelf life loss during that particular part of the freezer chain. After Kondrup, reproduced in Andersen, Jul and Riemann (1966). Ordinate is part of shelf life lost in one day at the temperature indicated.

and also used in the "Albany" test series, is discussed by the author in Jul and Dalhoff (1961).

The author (1960) gave examples of such calculations for ten different products. More up-to-date calculations are given below, cf. Tables 99-109 and Figs 127 - 145.

It is, of course, easy to construct a graph in such a way that estimates of loss of shelf life can be portrayed graphically. The ordinate in such a diagram is then, for each temperature used, the contribution to deterioration, i.e. shelf life loss which one day's exposure to that temperature will represent, i.e. the reverse of the acceptability time or shelf life in days at that temperature. Figure 77 gives an example of a complete freezer chain experience on the quality of a frozen product, quoted from Andersen, Jul and Riemann (1966).

As already mentioned, shelf life loss calculated in this manner has correlated well with that actually found when examining products after they had been subjected to the time-temperature experience in the freezer chain. Yet, further tests regarding the applicability of this method of calculating shelf life loss is an area which, in the view of the author, merits study.

When such calculations are carried out, it is important to stress that the figure at which one arrives, in the case illustrated in Table 67 the figure 0.220, represents degree of loss of shelf life. It does not necessarily mean that the product has lost 0.22 of its quality, cf. also van Arsdel (1969).

One may refer here to Fig. 32. Such products lose more of their quality during the first part of their storage life, while quality deteriorates slowly during the last period. Nevertheless, the product will have been exposed to such external influence which will cause it to loose a certain part, in the above case 0.22, of its total shelf life. A similar, but reverse situation, is illustrated in Fig. 33. Only where the quality loss is linear with time as in Fig. 31, can shelf life loss be equated with loss of quality, but this linearity of quality loss is by no means a necessary condition for the law of additivity to apply. The pork chops in the experiment which resulted in Fig. 33 would, after 400 days at -24°C, not have undergone any appreciable quality deterioration, yet they would have lost almost 2/3rd of their shelf life. Conversely, linearity of quality loss over time does not prove that losses during various time-temperature episodes are additive.

As indicated by Singh (1976), such calculations may easily be adapted for computer calculations. However, one should keep in mind that both TTT diagrams for shelf life and data on conditions in the freezer chain are not very accurate. Therefore, these calculations can be no more than indicative estimates, a fact often overlooked when sophisticated calculation methods are employed.

Components of the Freezer Chain

The freezer chain

To determine the effect of the freezer chain on the shelf life of a frozen product one needs, of course, to know the time-temperature conditions to which the food will be exposed. In the above example, Table 67, it was assumed that these conditions were known. The various stages of the freezer chain are different in different countries. For the USA as far back as 1952, Munter, Byrne and Dykstra (1953) suggested times and temperatures as indicated in Table 68.

McColloch *et al.* (1957) gave the times shown in Table 69.

Ronsivally (1981) suggests the maximum time a frozen seafood product should be held at -17.8°C in the various links of the freezer chain in Table 70.

Time-temperature surveys

As indicated above, it is reasonably easy to calculate an estimated end product quality, i.e. the quality as the user is likely to experience it, for a product entering the freezer chain, provided one knows the various time-temperature experiences to which the product may be exposed.

However, in day-to-day considerations of the performances of frozen foods during storage, transport, distribution and use, there is little advantage in being able to calculate a product's quality once its actual time-temperature experience is known, which at best would be at the time of end use, if at all. At that time a quality assessment could probably more easily and with greater accuracy be carried out by simply testing the product. Such a test would be more meaningful, also because the time-temperature experience of even an individual package is rarely known once a product reaches the ultimate consumer.

What is much more important for research workers as well as for the

Table 68. Time and temperature to which a product, on average, was exposed in the freezer chain in the USA about 1952, according to Munter, Byrne and Dykstra (1953).

Warehouse	–2.2°F	(–19.0°C)	for 6 months
Break up room	3.2°F	(–16.0°C)	for 1 week
Maximum during delivery	10.4°F	(–12.0°C)	for 12 hours
Backroom storage	3.2°F	(–16.0°C)	for 1 week

Table 69. Data regarding time spent in the frozen food chain in the USA. After McColloch *et al.* (1957).

Manufacturers' freezer store	30 days
Transport and transfer	10 days
Intermediate store	180 days
Retail store	14 days
Homes	7 days

Table 70. Maximum time that products should be held at -17.8°C in each of the five major elements in the chain of distribution of frozen fish fillets in the USA, according to Ronsivally (1981).

Vessel	7 days
Processing plant	1-2 days
Warehouse	6 months
Retail plant	3 months
Home	3 months

industrialists and the frozen food trade is to be able to determine beforehand, e.g. at the time of labelling, and with reasonable accuracy, what the quality is likely to be of a product at the time it is ultimately consumed and when it has been through the normal frozen food chain. Along with experience, such calculations would enable manufacturers to date label their products realistically and also indicate for them whether a product's keeping characteristics are sufficient for the market considered. Details of such calculations will, as illustrated in Table 67 and Fig. 77, also indicate where the most significant quality losses are likely to occur and, therefore, where vigilance or remedial action are most needed.

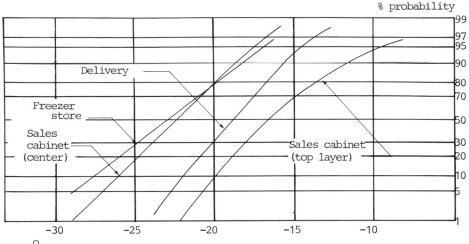

Fig. 78. Temperature distribution in the freezer chain in Denmark 1960. After Boldt (1961).

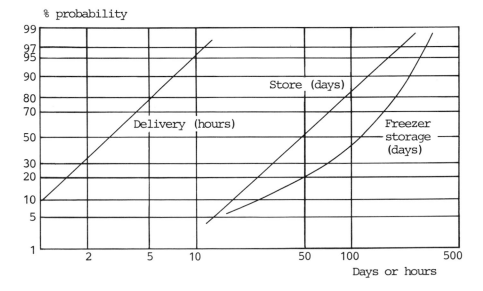

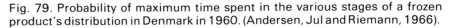

Fig. 79. Probability of maximum time spent in the various stages of a frozen product's distribution in Denmark in 1960. (Andersen, Jul and Riemann, 1966).

Table 71. Summary of conditions in the frozen food chain in Denmark, 1961.
(Andersen, Jul and Riemann, 1966).

| | Temperature °C | | Days | |
	50%	95%	50%	95%
Freezer store	−23	−17.3	120	285
Transport	−19	−15	0.12	0.4
Display, centre	−22.4	−17.7		
			50	180
Display, top layer	−16.8	−9.8		

Table 72. Estimate for mean times and temperature equivalents for the frozen
food chain in the USA. After Byrne and Dykstra (1969).

	Months in phase	Equivalent temperature
Primary warehouse	1.95	−23.3°C
Transportation and distribution	1.3	−16.1°C
In store and home	2.25	−13.3°C

This, however, suggests that one should know the time-temperature experience to which each product will be exposed. In principle, of course, this is impossible since individual shipments, cartons, and packages will receive different treatments. Nevertheless, it is highly relevant to carry out surveys of the frozen food chain to determine the probable time-temperature experience to which a product will be exposed.

To do this, surveys of time and temperature conditions throughout the freezer chain are required. Some such surveys were quoted above. Figures 78 and 79 indicate the data from a survey carried out in Denmark in 1960 by Boldt (1961).

From this it was suggested that, for the purpose of calculations such as those in Table 67, one use the data given in Table 71, i.e. conditions so adverse that 95% of the packages would not be exposed to them, that is, simply using the data from the 95% line in the probability diagrams. Of course, this was calculating with a very high degree of caution, since the chances of any one product being subject to the worst conditions in all cases are extremely minimal.

Corresponding estimates for the USA are given in Table 72.

The practical way of verifying the method of calculation would actually be to compare calculated qualities with the observed quality of products actually taken from display cabinets and placed in home freezers the appropriate length of time. This would be a somewhat complicated yet worthwhile exercise

% probability

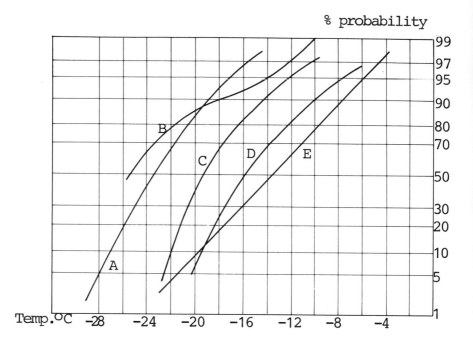

Fig. 80. Temperature in the freezer chain in Norway in 1970. A. Producer's cold store, B. Wholesale store, C. Long distance transport, D. Local distribution, and E. Freezer display cabinet. After Lorentzen (1971).

because it would eventually indicate the weight which one should assign to each link of the freezer chain. In other words, if one, using the 95% probability level, calculated that peas of a certain description would be below acceptability when consumed but in fact found most of the peas to be acceptable when purchased in the stores and even after home freezer storage one would clearly have used too pessimistic factors in the calculation.

Another survey of the freezer chain was carried out for Norway in 1970 by Lorentzen and co-workers; these data are given in Table 73, and Figs 80 and 81.

From this, G. Lorentzen, as quoted by Olsson and Bengtsson (1972), indicate that the chain in Scandinavia at that time was likely to be as illustrated in Table 73.

% probability

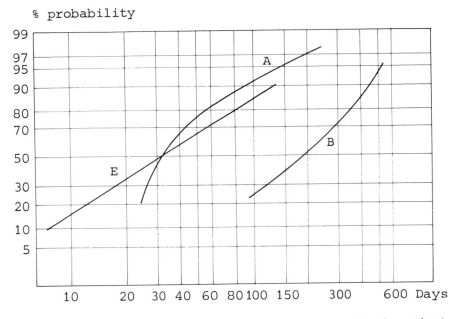

Fig. 81. Residence time for frozen foods in the various parts of the frozen food chain in Norway in 1970. Signatures explained under Fig. 80. After Lorentzen (1971).

Table 73. Data for the frozen food chain in Norway, measured by G. Lorentzen and quoted by Olsson and Bengtsson (1972).

| | Temperature °C | | Days | |
	50%	95%	50%	95%
Producer	−23	−17	40	150
Long distance transport	−19	−12	0 to 10	12
Intermediate storage	−25	−13	190	400
Local transport	−12	−8	1	7
Freezer display cabinet				
Top	−13	−7	74	100(90%)
Bottom	−19	−13		

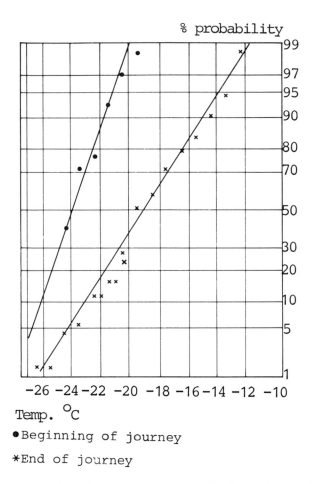

% probability

Temp. °C

• Beginning of journey

* End of journey

Fig. 82. Frequency of product temperatures at the beginning and the end of distribution journeys measured in different distribution trucks; mean ambient temperature: +7.9°C; average number of door openings per journey: 50. After Spiess, *et al.* (1977).

Spiess *et al.* (1977) gave some data for retail distribution in Germany, see Fig. 82.

A collaborative study carried out under the auspices of the IIR at the request of the Codex Committee on Frozen Foods gave results which for the Danish component of the study were given by Boegh-Soerensen and Bramsnaes (1977). Some of these results are summarized in Figs 83 and 84.

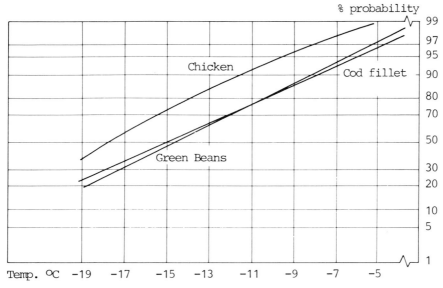
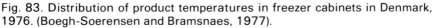

Fig. 83. Distribution of product temperatures in freezer cabinets in Denmark, 1976. (Boegh-Soerensen and Bramsnaes, 1977).

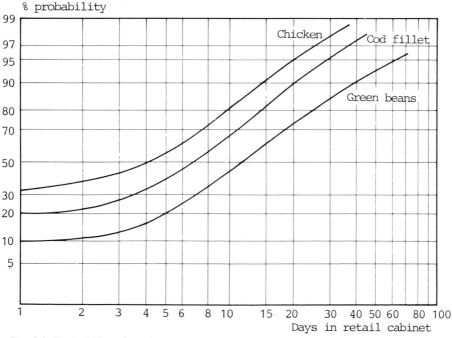

Fig. 84. Probability of products being stored in retail cabinets at times indicated or shorter in Denmark in 1976. (Boegh-Soerensen and Bramsnaes, 1977).

164   *The Quality of Frozen Foods*

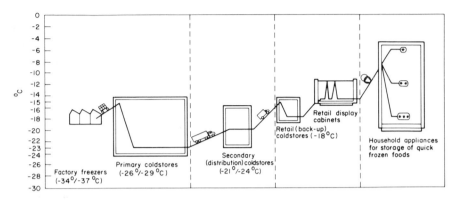

Fig. 85. Temperatures in the freezer chain in the UK as given by Sanderson-Walker (1979a).

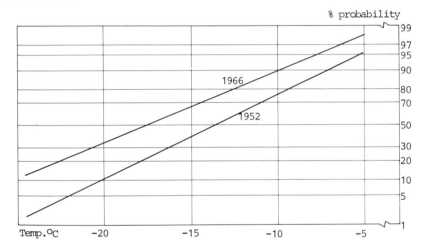

Fig. 86. Improvement in temperature of top layers in retail freezer cabinets in the USA. After Byrne and Dykstra (1969).

In so far as temperatures are concerned, M. Sanderson-Walker (1979a) suggested for the UK the temperatures given in Fig. 85.

Many of the data for surveys of time and temperature are from quite old observations. Byrne and Dykstra give the data for developments in temperatures in freezer cabinets reproduced in Fig. 86, and Lorentzen (1978) compared temperatures in Norwegian retail cabinets and arrived at the data shown in Fig. 87. A reduction in average temperature of 4.3°C was found over the period 1961 to 1977.

Lorentzen (1971) calls attention to the fact that giving average temperatures

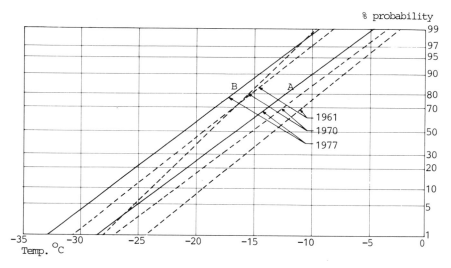

Fig. 87. Temperature distribution in freezer cabinets in Norway 1961, 1970 and 1977. A: top layers, B: diagonal point. After Lorentzen (1978).

Table 74. Average storage time for various frozen products in commercial cold storage warehouses, in months, in Sweden. After Persson and Nilson (1967).

Berries	5.5
Vegetables	5.5
Carcass meat	3.5
Cut up meat	1.5
Poultry	3
Fish	1.5
Herring	2
Bakery Goods	2.5
Ready cooked dishes	2

for, say, wholesale freezer stores will not adequately reflect the actual situation since the larger cold storage warehouses are likely to maintain the lowest temperatures. Thus, the major share of frozen foods may be exposed to lower temperatures than those assumed from some surveys. Thus, average temperatures - and storage times - should ideally be weighted according to capacity or turnover.

Residence times

Some early data from Sweden specifically related to storage times are given by Persson and Nilson and reproduced in Tables 74 and 75.

Table 75. Maximum time in days spent at wholesale and retail level for 90% of the products in the frozen food trade in Sweden. After Persson and Nilson (1967).

Berries	5.5
Vegetables	5.5
Carcass meat	3.5
Cut up meat	1.5
Poultry	3
Fish	1.5
Herring	2
Bakery goods	2.5
Ready cooked dishes	2

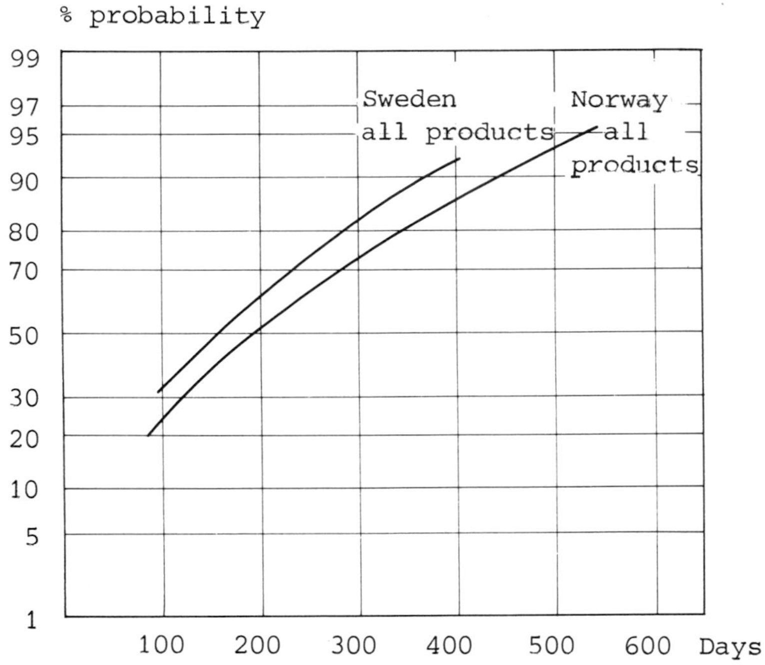

Fig. 88. Maximum residence time for frozen foods. Age judged with the aid of invoices in wholesale stores in Norway (1971) and in Sweden (1964 to 1967). After Olsson and Bengtsson (1972).

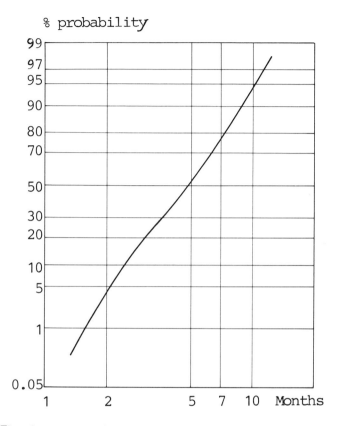

Fig. 89. Time between production and end use for five vegetables in the USA. Data collected through questionnaires placed in the packs. After Byrne and Dykstra (1969).

For the maximum residence time of frozen products at various stages, Olsson and Bengtsson gave the data reproduced in Fig. 88. Similar results from the USA are given in Fig. 89.

One characteristic which may be relatively easily measured is the average total time a product spends in the freezer chain up to the time of purchase. Thus, since marking with packaging data is mandatory in Denmark, Boegh-Soerensen and Bramsnaes (1977) obtained the results reproduced in Fig. 90.

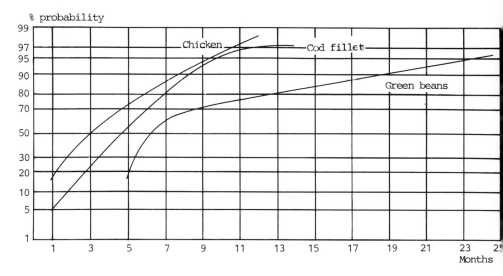

Fig. 90. Probability of products passing through the entire freezer chain in Denmark in time indicated or shorter. (Time from packaging to sale from display cabinet). After Boegh-Soerensen and Bramsnaes (1977).

Table 76. Times in weeks spent in sales cabinets in Norway, 1977, according to Lorentzen (1978).

Product group	Time (weeks)	Standard-deviation
Fish	10.4	14.1
Vegetables	11.8	17.6
Meat	8.7	11.7

Lorentzen (1978) observed that for Norway the average period a product spends in a sales cabinet is as indicated in Table 76, i.e. very large variations exist.

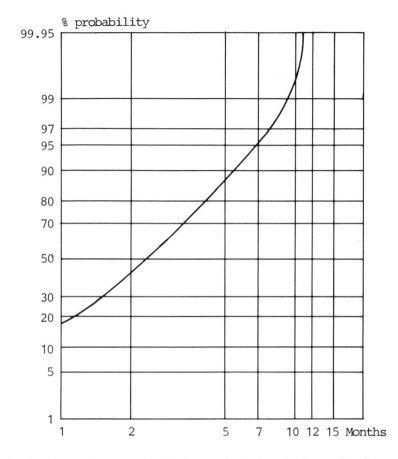

Fig. 91. Maximum time spent in the frozen food chain for frozen fish fingers in the Karlsruhe area. After Folkers and Spiess (1982).

For the Karlsruhe area in the Federal Republic of Germany, Folkers and Spiess (1982) arrived at the estimated total time in months from day of production to day of sale as indicated in Figs 91 - 93, i.e. more than 80% of fish fingers were sold less than 4 months after the day of production, for puff-pastry the corresponding time was 5 months, and for creamed spinach 10 months.

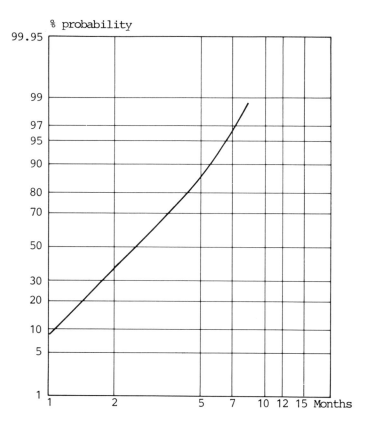

Fig. 92. Maximum time spent in the freezer chain in the Karlsruhe area for puff-pastry. After Folkers and Spiess (1982).

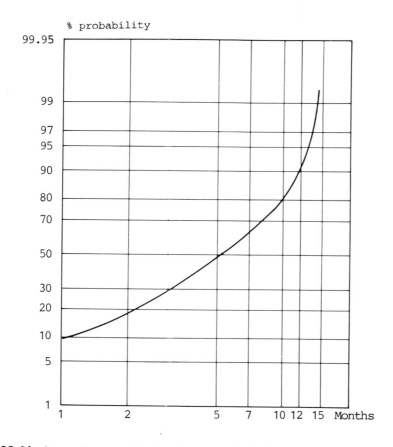

Fig. 93. Maximum time spent in the freezer chain for frozen creamed spinach in the Karlsruhe area. After Folkers and Spiess (1982).

They found the time spent as given in Table 77. They observed, also, that for date-marked products, no more than about 1% of all samples remained in the stores after the expiry of the date indicated.

Table 77. Percentile of observed age distribution of frozen products in the Karlsruhe area, observed in retail outlets. After Folkers and Spiess (1982).

Product	Maximum time (in months)		
	50 %	70 %	90 %
Fish fingers (5 pack)	3	4	6
Fish fingers (10 pack)	3	4	6
Fish fingers (15 pack)	3	3	4
Fillet of plaice	4	6	9
Königberger Klopse (meat balls with caper sauce)	4	5	6
Pizza	3	4	6
Puff-pastry	3	4	7
Creamed leeks	6	9	12
Creamed spinach	6	9	12
Green beans (chopped)	4	6	7

Table 78. Approximate turnover times for frozen foods in supermarkets in a large metropolitan area in the USA. Data from C.H. Byrne, 1966, quoted by Byrne and Dykstra (1969).

Product	Days
Peas	3.8
Spinach	5.2
Corn on cob	6.1
Broccoli spears	7.2
Fish sticks	7.8
"French fried potatoes"	8.5
Mixed fruit	9.8
Brussels sprouts	11.4
Raspberries	15.4
Strawberries	22.2
Peaches	55.6

For retail cabinets in the USA, Byrne and Dykstra gave the times indicated in Table 78.

In such surveys the reservations apply which also apply to marking with packaging date, i.e. date of packaging is often different from date of first freezing, etc. This is discussed later, but must be taken into consideration in this connection also.

In connection with the above mentioned collaborative work instigated by the Codex Committee on Frozen Foods, cf. Table 86 Boegh-Soerensen and Bramsnaes obtained some additional data reproduced in Figs 83 and 94 for

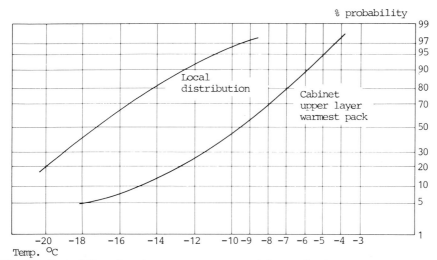

Fig. 94. Probability of various temperatures of frozen food on arrival at retail stores and of warmest package in top layer of various retail freezer cainets, measured in a survey in Denmark. After Boegh-Soerensen and Bramsnaes (1977).

Table 79. Typical food temperatures in some retail cabinets in one survey in Sydney. Defrost periods were not included. After Middlehurst, Richardson and Edwards (1972).

Store	Food	Top layer	Temperature in °C Second layer	Bottom
A	1 lb peas	−9 −18**	−16 −22	−27 −24
	1 lb chips	−8 to −0* −14	−15 −17	−26 −26
C	Chickens	−9 −12	−11 −12	−12 −11
F	Chickens	−13 to *+5		−13
	1 lb peas 1 lb peas	−11 −9	−13 −12	−15 −20 to −7

* Above load line
** Only 3 layers of packages in cabinet.

temperatures in retail cabinets.

For Australia, Middlehurst, Richardson and Edwards obtained the temperature data given in Table 79.

It is clear that additional studies of time-temperature in the freezer chain are indicated. Such studies should include the freezer chain all the way to consumption, i.e. normally the consumer's home since, as already mentioned, the quality of a frozen product in the retail cabinet should be such that it will remain fully acceptable at the time it is used by the consumer, i.e. often after a period of storage in the home freezer.

Cold storage warehouses and wholesale storage

Temperature management, i.e. temperature fluctuations and loading procedures in freezer storage warehouses, etc., will be discussed below. One aspect, however, deserves particular mention. It is always somewhat ideally stated that the "first in, first out" principle must be strictly adhered to in all parts of the freezer chain. However, not everybody who suggests this has actually tried to manage, let alone work, in freezer stores. These are often pretty fully loaded rooms with high stacks placed closely together often with different products packaged at different times. It is understandable that it is not easy to load and unload these stores with the degree of sophistication and accuracy which seems easy for a desk worker. It must be accepted that a fork truck operator, mainly looking for a pallet with a specified product and a specified brand name, may not get the product which arrived in the store first and which is often likely to be stacked behind more recent additions in the store room. It seems that only fully automatic handling systems in freezer storage rooms will overcome this problem. However, such systems are expensive, and not yet in widespread use. Some allowance for such human factors must be accepted.

Transport

Table 67 and Fig. 77 lead to some important conclusions with regard to the effects of the various links of the freezer chain. One factor, which normally receives much attention in discussions related to the quality of frozen foods, is the temperature in transport vehicles and variations during transit. Since most often the period of transport is relatively short, the effect of even quite warm temperatures during transport is rather negligible as will be seen from Fig. 77.

In this connection one may contemplate the considerable efforts which have been made to improve and control transport equipment for frozen foods. Most legislation is very strict about the temperature to be maintained during frozen food transport. Nationally, however, these requirements are rarely enforced. One case where enforcement is very strict is intra-European transports of perishable foods. Here, the Agreement on Transport of

Perishables, the so-called ATP-Convention, makes very strict requirements with regard to the quality of insulation, construction, etc., for vehicles transporting refrigerated or frozen foods. Also, the vehicles have to be tested and their insulating capacity verified at periodic intervals, and an international system has been built up for control and verification. This is interesting because, short of actually thawing, even major deviations from the prescribed temperature will not have any discernable effect on the quality of a frozen product transported. Yet, this particular piece of legislation has received more attention than any other regulation relating to the quality of foods, e.g. it has been ratified by 19 governments (1983), while four more are expected to do so in 1984.

The ATP-agreement includes a stipulation of temperatures to be maintained during transport; for frozen foods, these temperatures are for fish, fruit juice and products designated deepfrozen -18°C, for frozen foods in general -10°C except for offal, poultry and game -12°C. During defrosting, etc., a temporary temperature rise of 3°C is permitted. However, the Convention does not require recording of temperature during transport nor even the presence of a thermometer. One of the critical points of any such transport is that doors are often open for quite a long time during loading and unloading, custom formalities, etc. The author has witnessed a case where a shipment in ATP approved transport was held up for three days for custom formalities with doors open and refrigeration machinery turned off. Thus, a shipper is left with no guarantee as regards temperature during shipment, the only factor of importance for product quality. It appears that this is one example of the often experienced lack of sufficient communication between different disciplines.

While the technical rationale behind the ATP-Convention is not very clear, its very existence has speeded up considerably custom formalities for shipments of frozen foods and the above example appears to be an exception. It seems that the official nature of the system, the seal of approval on the vehicle, etc., has such a psychological effect that such shipments mostly are cleared without delay, regardless of quality and temperature of the goods.

At this point, it may be mentioned that the USA Code of Recommended Practices (The Frozen Food Roundtable, 1981) suggests that temperature be both measured in an appropriate place and recorded in vehicles used to transport frozen foods.

In assessing the effect of transport on quality, one must consider not only the transport vehicles but also, what is much more difficult to control, that is what happens at the points of loading and unloading. For instance, the Danish Meat Products Laboratory at one time found very well built refrigerated trucks picking up frozen poultry at a poultry slaughter plant. The product was taken right out of the freezer storage room and directly transferred to the trucks. However, at one point, the shipment was to go into

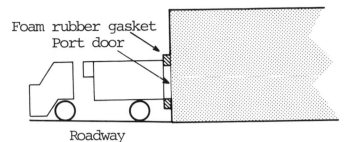

Fig. 95. Arrangement for transport from freezer store to freezer truck in contemporary design.

the freezer hold of a ship. Often, when the trucks arrived in the port, the ship had not yet arrived or was not ready for loading. The shipment was then just unloaded and left on the pier, at times even over night, before the pallets were transferred to the ship's hold.

A frequent occurrence is that a transport vehicle has to be emptied, because the most inaccessible part of the shipment has to be unloaded first. It is then even difficult to make certain that no undue delay occurs in placing the rest of the shipment back into the vehicle and maintaining the appropriate temperature in the latter.

Such experiences are by no means rare. This is yet another example where one controls with great care a factor which is easily controllable but leaves other factors, which are difficult to control but often more important, uncontrolled.

Similarly, Nicol and Spencer (1960) observed temperature rises in a product during loading of 2-4°C, much more than temperature rises in any ordinary mechanically refrigerated truck under adverse conditions. CRIOC (1982) reviews many cases where transport conditions deviate quite considerably from what is often assumed.

Modern cold store construction as suggested in Fig. 95 has done much to reduce temperature abuse at the point of loading and the wide acceptance of containerized transport has helped to minimize reloadings, etc. However, often only the cold store and primary part of the transport system is equipped for such care in handling, cf. also Meffert (1983).

In so far as the ATP-Convention is concerned, it is likely that it has benefitted the truckers involved more than the frozen food trade because the requirements and conditions are so that some truckers may have nearly a monopoly on international transport of frozen foods; rather high transport rates may have been a result. It is unlikely that the Convention has benefitted consumers at all.

Delaunay and Rosset (1981) indicate how transport vehicles for frozen food in France may be maintained either at -10°C or -20°C, according to the

Table 80. Specification for vehicles transporting frozen foods in France. After Delaunay and Rosset (1981).

Vehicle	Requirement	Ambient
Class B*	–10°C	30°C
Class C	–20°C	30°C
Isothermic**	Transports less than 100 km	

* Not for deep frozen products (surgelé)
** Only for a few frozen products (congelé), notably poultry.

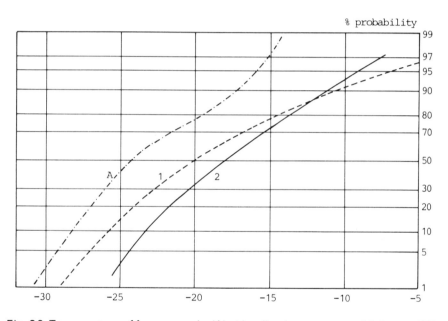

Fig. 96. Temperature of frozen goods, (1) at loading in transport vehicles, and (2) at unloading at new freezer store. (A) gives for comparison temperatures at originating freezer stores. After Lorentzen (1962).

product to be transported, cf. Table 80, and that, for shorter distances, isothermic vehicles may be used in some cases. These regulations appear to provide ample protection of quality. Ulrich (1981) concludes from a study of the extensive data available that temperature increases during transport have little effect on the quality of frozen fruits and vegetable products, except when they exceed about -6.5°C or are of considerable durations. One may well wonder why regulations for international transport in Europe need to be so

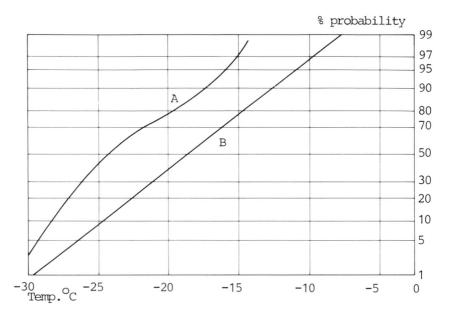

Fig. 97. Temperature at wholesale freezer store (A) and time of retail delivery (B). After Lorentzen (1962).

strict when regulations for transport in a country as vast as France can be more liberal, and may regret that ATP regulations could not have been based on a careful study of the extensive and readily available scientific literature on the effect on food quality of various temperature and time experiences.

It is worth noting that a recent draft proposal for a code of practice for the handling of frozen foods during transport, prepared for the Codex Alimentarius Programme (Codex Alimentarius Commission, 1983), contains quite liberal requirements and calls attention to the many weak links in the transport chain, mainly during loading and unloading. It proposes that a temperature rise of no more than 3°C over that recommended for the product be accepted.

Lorentzen (1962) gives in Fig. 96 interesting data regarding conditions for transporting frozen foods from producers to wholesale storage in Norway. As is seen, the products were exposed to considerably more heating-up prior to entering the transport vehicle than during transport. Figure 97 gives the corresponding data for transport locally to stores, where transport apparently is somewhat more critical.

In so far as transport is concerned, Ulrich (1981) indicates that Symons and Cutting (1977) refer to measurements in various European countries indicating that product temperatures after freezer transport were as indicated in Table 81. For Australia, Middlehurst, Richardson and Edwards (1972), see

Table 81. Product temperatures at end of deliveries of retail frozen food in various European countries, surveyed in 1972. After Symons and Cutting (1977).

Country	Temperature range
A	−23 to −9°C
B	−20 to −4°C
C	−21 to −5°C
D	−23 to −13°C
E	−18 to −16°C
F	−25 to −7°C
G	−21 to −4°C

Table 82. Temperatures of frozen foods during distribution to stores in one survey in Sydney. After Middlehurst, Richardson and Edwards (1972).

Food	Temperature in main warehouse cold store °C	Temperature on arrival at retail store °C
Chicken No. 6	−8*	−5
Chicken No. 8	−13*	−1
Chicken No. 9	−7*	−13
1 lb peas	−17	−12
1 lb peas	−17	−12
1 lb chips	−17	−9

* Left on loading dock for ½ hour.

Table 82, found temperatures as warm as -8°C at the time of loading at the freezer store and as warm as -1°C on arrival at the retail store, but do not report any adverse effect on quality.

It is the view of the author that this is an area where more detailed field studies are indicated. As mentioned, the construction details of road or rail vehicles, or even the way in which they are operated, are unlikely to be the weakest or even a weak link in the transport and transfer chain. Problems may arise from several loadings en route, failure or deliberate stops of the refrigerating machinery on the transport vehicle, whose noise often annoys a truck driver, etc. No testing of insulating data for a vehicle, etc., will disclose any of these factors, which in practice are of much greater importance. Foolproof temperature recorders with well placed sensors may disclose such abuses, albeit not prevent them. Ulrich (1981) points to the fact that most temperature abuse occurs, not in transport vehicles nor even in retail display cabinets, but during transfer.

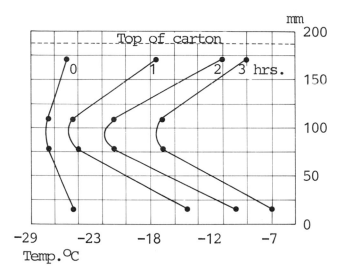

Fig. 98. Temperature rise at various depths of outer case in retail cartons with frozen fish, total weight 9.5 kg, placed on concrete floor at 21°C ambient temperature. After Nicol and Spencer (1960).

One development which may be found useful for monitoring during transport is that of the various time-temperature recorders and integrators discussed later. A very simple system is described by Blixt (1983). It consists of a very flat, almost label like, inexpensive indicator which, by an irreversible colour change, will show if a certain preset temperature has been exceeded.

Over and above this, it is worth recalling the data in Table 67 and Fig. 77. Shelf life loss during transport and transfers are normally insignificant, even in cases of some considerable abuse.

Product handling

Nicol and Spencer (1960) measured temperatures in consumer packs of frozen fish in an outer carton, weighing 9.5 kg, when placed on a concrete floor at an ambient temperature of 21°C, as shown in Fig. 98.

It can be seen that a few hours inadvertently spent under these conditions leads to significant temperature increases. Close surveillance of the loading and unloading operations, especially at the retail level, may thus be much more important than verifying the conditions of the transport vehicles. In studying a large vehicle with about 13 cm of wall insulation and a mechanical refrigeration unit, Nicol and Spencer found maximum temperature rises of 2.2 to 4.4°C in 24 hours.

Table 83. Probability, in per cent of time spent, in carrying frozen foods home from two department stores in Sweden. After Olsson and Bengtsson (1972).

Min.	Store A	Store B
0-15	48	52
15-30	35	34
30-45	9	7
45-60	4	3
60	4	4

There is much to indicate that much could be achieved by more vigilance and possibly some regulations regarding the handling of shipments of frozen foods which often (as mentioned above) are left on piers, loading platforms, etc., for some undefined period of time. Data developed by the Danish Meat Products Laboratory (1981) indicate that in a pallet of cartons with frozen broilers, with an initial temperature of -20°C, the temperature had increased to -10°C in the top layer after 3 hours if no protective cover was provided. If an ordinary tarpaulin were used to cover the palletized goods the products did not fare much better; after 3 hours the top layer was at -12°C. However, if a cover of heat reflecting material was used, the top layer did not undergo any measurable temperature increase from the original temperature of -20°C during the three hours out of refrigeration. Ambient temperature was 19°C. Also this field is in need of further studies which could then be the base for official regulations, if such are indicated.

Löndahl (1977) found that the center temperature of an outer pack of spinach loaded on pallets increased from -25°C to -10°C in one hour at ambient temperatures of +15°C.

Delaunay and Rosset (1981) indicate how, in France, guidelines are given for handling, loading and unloading frozen products so as to avoid abuses, such as the ones described above. Middlehurst, Richardson and Edwards (1972) noted that one weak link in the freezer chain was the process of loading freezer cabinets in the stores. In their studies, frozen products were often kept outside the cabinets in outer cartons for considerable periods before they were placed in the cabinet. Temperature abuse at the time of loading into trucks at the main warehouse or at unloading at the store was also common as is illustrated in Table 82.

Another weak link in the chain is the period after individual packages are brought from the shop to the consumer's home. This may simply be considered part of the thawing process if the product is used immediately. If the product is placed in a home freezer, however, this period may have some undesired effect. Table 83 reproduces a study by Djupfrysningsbyraan, Sweden (1970), and quoted by Olsson and Bengtsson (1972), related to this aspect.

Table 84. The warming-up of frozen food packages if left unprotected, e.g. during transport home. Data from G. Strandell, quoted by Olsson and Bengtsson (1972).

Product	Time, hours	Temperature, °C Surface	Centre	Surrounding temperature
Cod fillet	½	−6	−6	5°C
	1	−4	−5	5°C
(450 grams, packed in	½	−3	−6	20°C
waxed carton)	1	−2	−4	20°C
Broiler	½	−8	−10	5°C
	1	−6	−9	5°C
(1.26 kilos	½	−5	−10	20°C
Cryovac)	1	−3	−6	20°C
Chopped spinach	½	−3	−4	20°C
	1	−2	−3	20°C
(600 grams, waxed carton)				

A study by G. Strandell, quoted by Olsson and Bengtsson (1972), suggests that if frozen foods are -20°C at the time they are taken from the sales cabinet, they will have the temperatures indicated in Table 84 at the time they reach the consumer's home.

Sales cabinets

Temperatures in sales cabinets have been the subject of a great many studies over the last 20 years. Table 85 and Figs 83 and 87 show that there may be a very wide variation beween the temperatures of various frozen food packages in retail freezer cabinets. Sanderson-Walker (1979b) mentions that, in the UK, 8% of the products were found warmer than -12°C, and 65% colder than -15°C.

On the whole it is likely that a not inconsiderable part of the total quality loss, which a retail packaged frozen food product suffers, takes place in the retail cabinets. One interesting fact emanates from the time-temperature surveys carried out by Boegh-Soerensen and Bramsnaes (1977), see Fig. 84, and those carried out by Lorentzen (1978), see Fig. 81. It seems as if the average time a product is kept in a sales cabinet is considerably shorter than some years ago, i.e. on average 8 days as indicated in the study by Boegh-Soerensen and Bramsnaes for 1976 as compared to 50 days in 1960 as indicated in Fig. 79. The reason for this is probably not a concern by store management for the quality of the products but the obvious fact that the

Table 85. Observations of temperatures in retail freezer cabinets of the open-top type in France. After Comité Interministériel de l'Agriculture et de l'Alimentation (1980).

	Surface			Centre			Bottom		
	Green beans	Fish	Poul- try	Green beans	Fish	Poul- try	Green beans	Fish	Poul- try
−18°C or below	101 (33.9%)	146 (0.4%)	4 (16.7%)	184 (62.2%)	240 (54.8%)	13 (59.1%)	229 (75.6%)	312 (73.2%)	23 (85.2%)
Between −15 and −17.9°C	77 (25.8%)	109 (24.9%)	10 (41.7%)	50 (16.9%)	102 (25.3%)	5 (27.7%)	35 (11.5%)	58 (13.6%)	3 (11.1%)
Between −12 and −14.9°C	62 (20.8%)	97 (22.4%)	3 (12.5%)	36 (12.2%)	62 (14.1%)	3 (13.6%)	25 (8.3%)	42 (9.9%)	1 (3.7%)
Above −12°C	58 (19.5%)	85 (19.5%)	7 (29.1%)	26 (8.7%)	34 (7.8%)	1 (4.6%)	14 (4.6%)	14 (3.3%)	— —

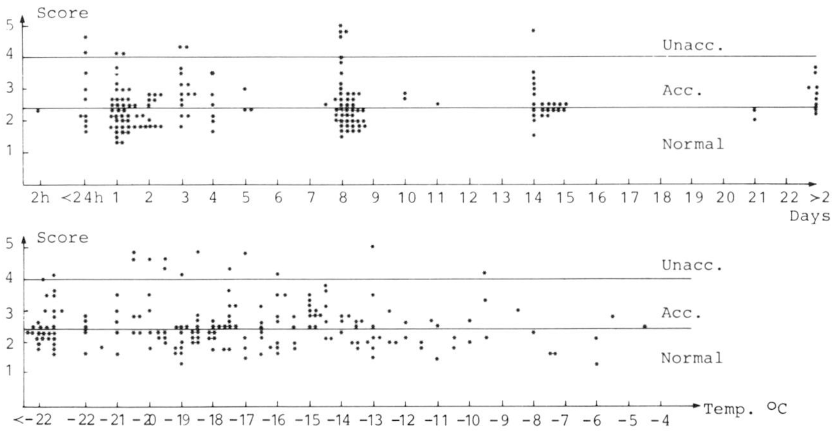

Fig. 99. Quality score plotted against estimated length of time in cabinet, and temperature of frozen chicken at time of sampling for products included in survey illustrated in Figs 83 and 84. Data received from Boegh-Soerensen, Denmark.

stores have a great economic interest in a rapid turnover and quickly discontinue carrying product lines which move slowly.

Ronsivalli (1981) indicates that frozen fish fillets in the USA in most cases are held less than 5 days in the retail cabinet.

Hammer (1976) reviewed the deliberations of the Codex Group of Experts on Frozen Foods up to 1976. He indicates that, in the light of the temperatures

Table 86. Quality of products in Codex Alimentarius instigated investigation. Codex (1978)

	Normal A	Acceptable B	Unacceptable C	Total D
Total				
Samples	1727 (73.5%)	504 (21.4%)	119 (5.1%)	2350
Product				
Chicken	619 (84.3%)	99 (13.5%)	16 (2.2%)	734 (31.2%)
Green beans	530 (63.7%)	240 (28.8%)	62 (7.5%)	832 (35.4%)
Cod	578 (73.7%)	165 (21.1%)	41 (5.2%)	724 (33.4)%)
Type of store				
Non-self-service	249 (62.1%)	109 (27.2%)	43 (10.7%)	401 (17.1%)
Self-service	1478 (75.8%)	395 (20.3%)	76 (3.9%)	1949 (82.9%)
Condition of cabinet				
Reasonably maintained	1562 (73.6%)	454 (21.4%)	105 (5.0%)	2121 (90.3%)
Poorly maintained	165 (72.1%)	50 (21.8%)	14 (6.1%)	229 (9.7%)
Position in cabinet				
Above loadline	101 (67.9%)	45 (27.8%)	7 (4.3%)	162 (6.9%)
Below loadline	1617 (73.9%)	459 (21.0%)	112 (5.1%)	2188 (93.1%)
Temperature				
−11.99°C and warmer	614 (72.5%)	193 (22.8%)	40 (4.7%)	847 (36.0%)
−12°C and colder	1113 (74.0%)	311 (20.7%)	79 (5.3%)	1503 (64.0%)
of which				
−12°C to −17.9°C	665 (72.8%)	191 (20.1%)	57 (6.3%)	913 (38.9%)
−18°C and colder	448 (75.9%)	120 (20.4%)	22 (3.7%)	590 (25.1%)

Note: Percentages in cols. A-C refer to the total for that line; shown in col.D, percentages refer to total samples.

observed in retail cabinets in surveys in various countries revealing a high incidence of quite warm temperatures, the Group did not consider it appropriate to make proposals regarding temperatures in this part of the freezer chain.

One other fact emerged from the above mentioned study by Bramsnaes and Boegh-Soerensen (1977). As illustrated in Table 86 and Fig. 99 there did not seem to be any correlation between the organoleptic quality of the products when they were taken from the cabinets and their time therein nor with their temperature at the time of withdrawal. When the random nature of each product's temperature at withdrawal and of its experience prior to withdrawal and testing is considered, it is obvious that it would be so and this fact cannot, as is sometimes done, be taken to indicate that the temperatures of the products in retail cabinets have no effect on product quality.

McBride and Richardson (1979) pointed out that experimental data do not support the general requirement that frozen foods should be distributed at temperatures as cold as -18°C at all levels.

J. Philippon, France, describes how tests such as these shown in Table 85 are carried out in France every year and summarized according to the various types of retail outlets. The results are made available to the various groups of stores, discussed with store managers, etc. This is said to create a general awareness of this problem area and to be a considerable motivation to improve.

One factor, in addition to the temperature in the cabinet, is that the products are exposed to considerable handling and shifting around. Even puncturing of packages, and other abuse, may occur because of the customers' usual way of selecting products in a self-service store. In addition, inherent in the construction of gondola type display cabinets, is the fact that the temperature cannot be the same throughout the cabinet; a considerable temperature gradient will exist. Fluctuations may also be due to defrosting, to consumers' moving the packages around and because the stock is frequently replenished with new products. All this results in very considerable fluctuations of the temperature to which a package is exposed.

The temperatures measured in retail cabinets in various surveys conflict with the conventional belief that frozen products can and should be held at a very cold temperature in freezer display cabinets, in Europe often put at -20°C with -18 or -15°C as a tolerated, exceptional and occasional maximum. This belief has even been translated into official rules in many countries, cf. CRIOC (1982). The fact that it is not adhered to seems generally to have been conveniently overlooked. Because of much attention recently paid to this, it has frequently, even in EEC contexts, been suggested that the requirements, i.e. -20°C with a maximum of -18°C, should now be enforced and, what would then be termed as defective cabinets, should be replaced by some equipped to maintain the specified temperature, probably allowing for a lead time of some years before replacement was required. One might even anticipate strong support for such a move from equipment manufacturers. As mentioned above, in a new draft directive (1982), the Commission of the European Communities has suggested a rather liberal maximum temperature, i.e. -18°C with a tolerance of 6°C, i.e. basically accepting prevailing conditions.

In this context, an article by Middlehurst, Richardson and Edwards (1972) makes interesting reading. It describes how the discovery that store freezer cabinets in Australia did not all conform to official requirements caused some immediate concern but became accepted as unavoidable. It was also realized that the situation did not result in frozen foods being substandard and it was gradually accepted. However, the exercise led to some very useful recommendations which may have done more to improve product quality than a courageous attempt to have all freezer cabinets replaced would have done. Thus, it was recommended that (i) no product should ever be above the load line, (ii) that a night cover should be used, (iii) frozen foods should be placed in the freezer cabinet immediately on arrival at the store unless back room freezer storage was provided and products should not be permitted to stand in outer cartons until placed in the freezer cabinet, and (iv) rapid turnover and a

Table 87. Comparative refrigeration performance in W, at various condensor and evaporator temperatures, for a small compressor with a 4 kW motor, at 20°C. Data from Joergen Lorentzen, (1981).

Condensor temp.	Compressor efficiency in W 4 kW motor		
	30°C	40°C	50°C
Evapor temp. °C	Performance, W		
−5	17 150	15 360	13 860
−10	14 040	12 540	11 270
−15	11 440	10 130	9 140
−20	9 140	7 980	7 160
−25	7 080	6 120	5 570
−30	5 520	4 720	4 260
−35	4 160	3 460	3 140
−40	2 980	2 420	2 190
−45	2 010	1 620	—

"first-in, first-out" system should be adhered to.

Persson and Nilsson (1967) similarly suggested that retail stores ought to be better equipped with freezer storage rooms for receipt of shipments of frozen foods.

Energy and retail cabinets

As far as some regulating authorities are concerned, there still seems to exist a determination to enforce the maintenance of cold temperatures, i.e. -18°C or below, in any part of the loading volume of freezer sale cabinets. This requirement is also supported in UKAFFP (1978).

However, it takes only a little engineering knowledge to realize that implementing a rule that sales cabinets be equipped to maintain -20°C or colder throughout would be a very costly exercise. It would involve replacing all existing gondola (open-top) type freezer cabinets. The cabinets would have to be of a much more expensive design than the existing types, so much so that it would be a considerable additional cost in distributing frozen foods. It would require reducing the temperature in the cabinets by approximately -8°C. Thus evaporator temperature, presently about -30°C, would have to be lowered. If such arrangements were feasible at all, this might mean that an extra stage would have to be added to the compressors for the cabinets.

The energy consumption per Joule to be removed would increase dramatically due to the lower efficiency of the refrigerator equipment at the lower temperature. This is seen from Tables 87 which shows the relative energy

Table 88. Energy consumption in the various links of the freezer chain in France. After Livsmedelsteknik (1980).

	kWh per MT	1000 MT per year	Total 10^6 kWh per year
Freezing	150	500	75
Freezer storage	200	600	120
Freezer transport	16	600	9
Freezer sales cabinets	1000	240	240
Home freezers	5250	270	708

removal for a certain compressor type with a 4 kW motor at various evaporator temperatures, cf. also Fig. 19. Reducing temperatures in all parts of a retail freezer cabinet as suggested, i.e. to be not warmer than -20°C in any place, is estimated from these factors to result in a 1.5-fold increase in energy requirement. In addition, however, the lowered temperature would also result in higher heat penetration into the cabinets, to be offset by additional compressor capacity, i.e. if the present temperature differential is about +20-(-15)°C = 35°C, the new temperature differential would be increased to +20-(-25)°C = 45°C or by a factor of 1.3 for a regular commercial display cabinet. Thus, in total resulting energy requirements in the retail chain could be increased by a factor of 2.

In quoting comparisons between home freezers operated at -12°C and -20°C, Meffert (1983) arrives at somewhat similar conclusions regarding energy consumption at these two temperatures.

Therefore, in considerations regarding lowering temperature in sales cabinets and in order to avoid exhorbitantly wasteful requirements, some sort of condensor-heat utilization would have to be arranged, as is already the case in some stores where heat from condensors is used for space or water heating. Even in colder climates and in cold seasons there would not be a need for so much energy for space heating. Heating water might be a better possibility, but there might also not be a need for these amounts of warm water. In such utilization, the efficiency would hardly ever exceed 70%. Thus, steps to reduce temperatures in conventional sales cabinets would, even at best, lead to high increases in energy consumption.

The order of magnitude of energy requirement of freezer sales cabinets is estimated to be around 0.6 kWh per m of length of cabinet per hour, as indicated by Jes Brinch, Denmark, i.e. a very considerable part of a store's total electricity consumption.

Fine (1981) estimates that the requirements of the food refrigeration systems constitute 55% of the electricity requirement of a normal supermarket.

Table 88 indicates the relative energy consumption in the frozen food chain

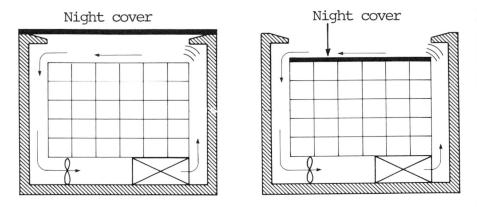

Fig. 100. Two methods of using night covers. After Boegh-Soerensen and Bramsnaes (1977).

determined in a survey in France. It shows that by far the largest energy input in the commercial sector is for maintaining low temperatures in retail cabinets, obviously because of the relatively very large heat absorption which takes place in these small units with a low volume-to-surface ratio compared to freezer storage warehouses or even freezer vehicles. Thus, a decision to maintain -20°C or colder throughout sales cabinets, which often just seems to mean implementing existing regulations, would have very serious consequences for economy and energy use.

Therefore, before any such step is even contemplated, one should seriously consider the advantages which would be derived from such a step. It appears that frozen foods, as presently available to consumers, are perfectly acceptable. Even if the period spent by products in retail cabinets represents a quality loss, it has limited effect on total quality, cf. Table 86 and Fig. 99 and also Table 67. Therefore, an enormous cost would be the result and the benefit derived would be negligible and likely to remain completely unobserved by consumers, whose satisfaction would really be the purpose of the whole undertaking.

Improved cabinet management

One simple improvement could be the use of night covers on all present self-service cabinets when the shop is closed, as discussed by Boegh-Soerensen and Bramsnaes (1977), cf. Fig. 100. Figure 101 gives temperatures in a retail cabinet (open chest) with and without lid. Eskilson (1979) points out that where night covers are placed directly on the products, its upside should be reflecting material. Conversely, where night covers are placed above the circulating cold air, its underside should be reflecting material.

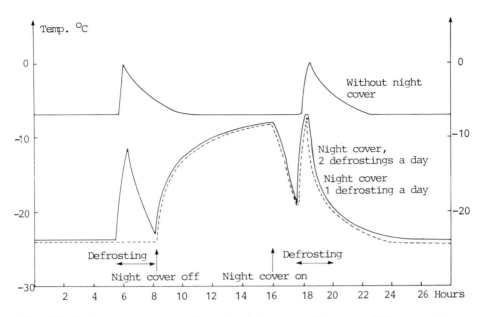

Fig. 101. Surface temperature of product in freezer cabinet over 24 hours, with and without night cover. — two defrostings per day, ... one defrosting per day. After Boegh-Soerensen and Bramsnaes (1977).

Table 89. Frost formation (in % of net weight) during storage of meat balls in the upper layer of a freezer cabinet. Where night cover was used, it was during non-selling periods. After Brinch, Boegh-Soerensen and Else Green (1982).

	Cabinet time (days)			
	10	20	30	40
Night cover	3.1	7.5	5.6	3.8
Without night cover	5.8	12.3	14.9	10.8
Control (–30°C)	1.4	0	0.7	1.6

The average product temperature where night covers are used is much lower than without the lid. However, it must be noted that the product in the former cabinet will experience a much greater temperature fluctuation than in the other cabinet. It then becomes necessary to determine whether this temperature fluctuation might possibly offset the advantages of the lower night temperature, see also Meffert (1983). The data reproduced in Tables 89 - 91 show that the use of night covers resulted in less in-package frost formation and a better retention of overall quality when the samples were presented to a taste panel.

Table 90. Taste panel evaluation of same product as in Table 89. After Brinch, Boegh-Soerensen and Else Green (1982); (Score: -5 dislike extremely, +5 like extremely).

| | Cabinet time (days) | | | |
	10	20	30	40
Night cover	1.5	1.2	1.1	1.0
Without night cover	1.1	0.9	1.0	0.4
Control (−30°C)	1.4	1.4	1.0	1.1

Table 91. Frost formation in packs with frozen peas in open-top cabinets with forced air circulation. After Brinch, Boegh-Soerensen and Else Green (1982).

| | Plastic pouch | |
	PE	Alu-laminate
Night cover	6 %	4 %
Without cover	11 %	5 %
Control (−20°C)	2 %	

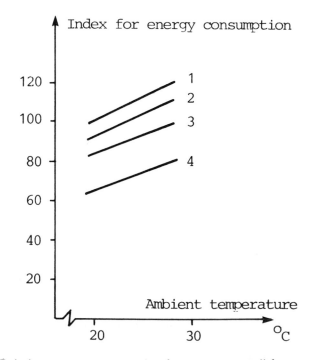

Fig. 102. Relative energy consumption for open-top retail freezer cabinets by regular operation (1), with night lid directly on packages (2), with night lid covering cabinet top (3), and compresser switched off nights, with only one defrosting daily (4). After Brinch, Boegh-Soerensen and Else Green (1982).

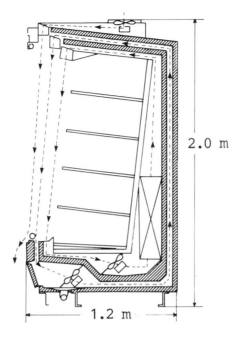

2.0 m

1.2 m

Fig. 103. Vertical freezer display cabinet (multi-shelf open desk cabinet). For this construction also, night covers may be used. After Eskilson (1978).

Brinch, Boegh-Soerensen and Else Green showed that the use of night covers may also result in significant energy savings, see Fig. 102.

Hawkins, Pearson and Raynor (1973) showed that the use of reflecting night blinds may have a very beneficial effect on product quality, cf. Fig. 104. These refer to the visual quality of the product. Differences for cooked products were not nearly as striking, yet the author considered them sufficient to make the use of night blinds worthwhile, at least if more than one to two weeks in the cabinets is to be expected.

One aspect stressed particularly by Middlehurst, Richardson and Edwards (1972) is that systematic stock rotation and maintenance of the above mentioned first-in first-out principle may do more to products' quality than very costly cabinet modifications. Similarly, loading procedures should be supervised and keeping frozen foods unprotected temporarily outside of freezer cabinets or freezer stores should not be permitted, as mentioned earlier.

Löndahl (1977) measured the center temperature of the outer pack in a master carton containing ten consumer packs of spinach exposed to ambient temperature of 15°C, approximating conditions in a store. It increased from

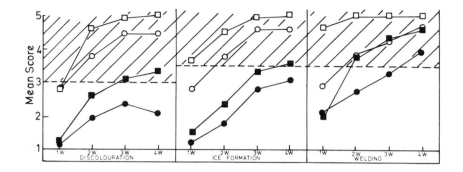

Fig. 104. Results of subjective assessment of visual quality of peas in polybags. (Score: 1 = no defects; 5 very severe defects); O convection cabinet. ● convection cabinet with night blind; □ forced air cabinet; ■ forced air cabinet with night blind. After Hawkins, Pearson and Raynor (1973). Scores in shaded areas are considered not acceptable. Time is indicated in weeks (W).

-25°C to -10°C within an hour.

Middlehurst, Richardson and Edwards (1972) also observed that it is very important that frozen foods are never stored above the load line.

Blixt (1983) reports on developments by the I-Point corporation in Malmö i.e. a small and very inexpensive self-adhesive, label-like tag which may be placed in various places of a frozen food cabinet. Its colour will indicate if one of several predetermined temperatures is being exceeded in that spot. The colour change is reversible, i.e. the colour will change back to normal if temperatures around the indicator are lowered sufficiently.

It might be better to try to improve designs of sales cabinets, conceivably considering using some with easy opening, automatically closing, frost free and completely transparent lids, vertical cabinets with glass doors or transparent curtains, etc., rather than making further attempts to pass stricter legislation.

One other feature worthy of study would be means whereby the hot defrosting air is not fed over the top of the merchandise as is presently the case in gondola-type cabinets.

Especially because of energy considerations, there is today a definite trend away from the open-top gondola-type cabinets towards vertical cabinets with glass doors, cf. Figs 105 and 106. One problem is that placing products frozen in plastic bags gives special difficulties in such cabinets. Also, there may be a tendency to install excessive lighting in vertical cabinets, causing too much

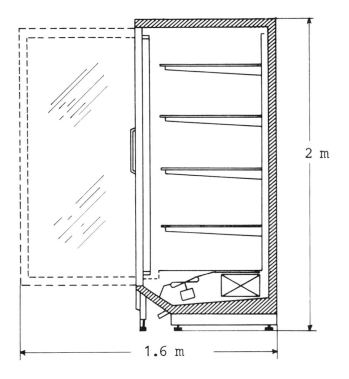

Fig. 105. Vertical, closed door type freezer display cabinet (glass door case). Data from Leif Boegh-Soerensen (1982b).

heat absorption by the packages. It is fortunate that the tendency is to use the gondola-type sales cabinets mainly for products with a rapid turnover.

Boström (1982) describes vertical retail sales cabinets. He estimates that vertical cabinets for a certain sales volume require 9.5 m² floor area. For the same sales volume he estimates that horizontal, gondola-type cabinets require 28.8 m², cf. Fig. 106. He estimated that there were about 240 such vertical cabinet installations in Scandinavia (1982). He suggests that such cabinets may maintain a fairly constant and uniform temperature down to -23°C throughout the loading volume.

Where cabinets can be arranged so that they can be filled-up from behind there is the additional advantage of easy adherence to the first-in- first-out principle.

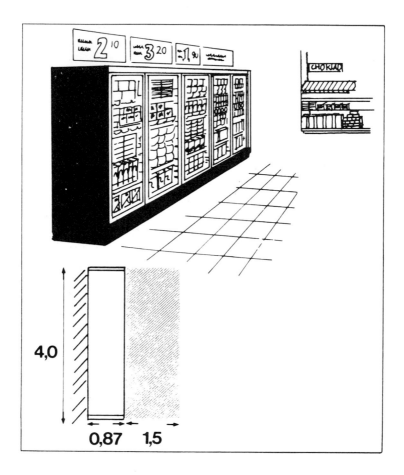

Fig. 106. Comparison of vertical and horizontal retail cabinets both with a 2.8 m³ capacity for products. Measures in metres. The vertical cabinets requires 9.5 m² floor space while the horizontal, gondola-type requires 28.8 m². After Boström (1982).

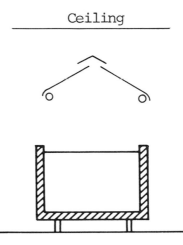

Fig. 107. Deflectors for heat rays over retail cabinet to reduce infra-red radiation. After Delaunay and Rosset (1981).

Delaunay and Rosset (1981) indicate how, in France, various measures are proposed or prescribed for the management of sales cabinets in retail stores, e.g. moderate light intensity, obtained by fluorescent light, may not exceed 600 lux, air drafts should be avoided, etc. Installations, such as that shown in Fig. 107, are recommended. To be effective, such a deflector must be placed very low over the cabinet (Eskilson, 1983).

Hawkins, Pearson and Raynor (1973) have studied the effect of three steps to reduce product surface temperatures in the top layer of an open-top frozen food display cabinet.

First, they tested the use of highly reflective aluminium foil with an emissivity of 0.05 in comparison with foil with yellow lacquer, emissivity 0.30 and white paper or waxed paper (emissivity 0.84 to 0.90). They realized that completely reflecting foils are unlikely to be used but found that with some printing, etc., one may obtain an emissivity of 0.30 and a temperature reduction of 10°C of the most exposed packages.

Hawkins, Pearson and Raynor (1973) also tried using reflecting night blinds. In a forced air cabinet, these were very effective in reducing average temperature. However, peak temperatures during defrosting remained practically unaffected.

They also tried placing reflecting mirrors, placed much like the two reflectors shown in Fig. 107. This gave a temperature reduction of about 3°C.

As the same authors mention, a similar effect may be obtained by the use of corner-cube reflectors, a special highly reflecting canopy with a number of indentations such as by the corner of a cube pressed against the sheet.

Table 92. Temperatures observed in home freezers in the USA about 1962, quoted by Olsson and Bengtsson (1972).

Temperature, °C	%
Below −18	30
−18 to −12	37
−12 to −7	22
Above −7	11

Table 93. Temperatures in home freezers in the USA, observed by Redstrom (1971).

Temperature, °C	%
Colder than −18	65
Warmer than −12	6

Detailed advice is also given regarding procedures for placing packages in the sales cabinets, loading the cabinets and close monitoring of performance of both equipment and personnel. Where no separate freezer storage rooms are available at the retail level, products may be maintained temporarily in an isothermic container until placed in the sales cabinet.

One should recall also that changes in heat radiation reflectivity characteristics of the packaging material could decrease temperatures in the surface layers of frozen food packages as much as 8°C as measured by Dykstra, quoted by Olsson and Bengtsson (1972), and also summarized above.

Observations, such as the above, caused Middlehurst, Richardson and Edwards (1972) to propose that the maximum permissible temperature anywhere in a frozen foods sales cabinet could be 15°F, (-10°C).

Any steps to enforce improvements in retail frozen food management need to be viewed from the standpoint of the quality improvement which would actually be experienced by the consumer.

Home storage

Common to practically all time-temperature surveys is that a storage period in the consumer's home, i.e. in a home freezer, has been left out. Such a period should probably be included in estimates regarding quality losses in the frozen food chain. Some early data for the USA are given by Olsson and Bengtsson (1972), cf. Table 92.

Redstrom (1971) gave the data for conditions in home freezers in the USA, reproduced in Tables 93 - 95. These data include home frozen products and should be viewed with that in mind.

Table 94. Data on temperatures and storage times in home freezers in the USA, reported by Redstrom (1971).

	Country	Town
Home freezers with temperature warmer than −18°C	65%	60%
Food stored less than 1 week	25%	25%
Food stored less than 2 weeks	-	50%
Food stored less than 3 weeks	50%	-
Food stored less than 180 days	-	95%
Food stored less than 300 days	95%	-
% of fruit and vegetables left after 7 months	31-44%	11-27%

Table 95. Periods products are kept in home freezers in the USA, reported by Redstrom (1971).

Used within	%
1 week	25
3 weeks	50
30 weeks	95

Sharp and Irving (1976) also found that commercially produced products were kept in home freezers for the periods and at the temperatures indicated in Figs 108 - 109 before they were actually used. Their data are further illustrated in Figs 110 and 111. The illustrations indicate that at least sensitive products may suffer quite a substantial additional shelf life loss in home storage before their use.

Future studies regarding the keeping characteristics of frozen food during distribution should include this aspect since it seems obvious that a frozen food producer would want the products to be capable of absorbing some shelf life loss in a home freezer without becoming unacceptable. Thus, as Couden (1969) points out, industry normally operates on a so-called "margin concept", i.e. some shelf life time should be retained providing for possible

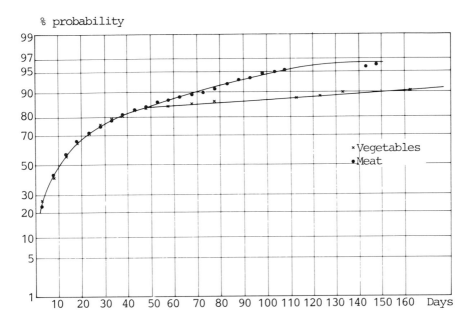

Fig. 108. Average number of purchased frozen food packages withdrawn from home freezers within five-day time intervals for two categories of commonly stored food: medium-sized meat, such as chops and steak, and vegetables other than potatoes and peas, according to Sharp and Irving (1976).

storage time in the consumer's home. Likewise, a consumer would demand that this be the case. Therefore, any prediction of a product's quality at the time of actual use should take into account the likelihood that the product will undergo such an episode.

It is likely that some such studies of time in home freezers have a built-in bias if used in calculations such as those exemplified in Table 67 in that frozen foods may be brought home for immediate thawing and consumption; thus, their fate may not be included in such studies. On the other hand, Jytte Kjaergaard, Denmark (1983), indicates from her own surveys that people rarely buy frozen foods, other than ice cream and fish, for immediate use.

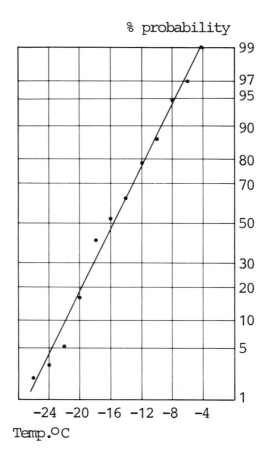

Fig. 109. Probability of households in Australia in which the temperature of the frozen food compartment of the refrigerator or the freezer was lower than or at temperature indicated, according to Sharp and Irving (1976).

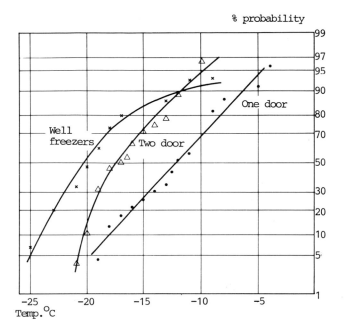

Fig. 110. Average storage temperature in home freezers according to different freezer types, in Australia, according to Sharp and Irving (1976).

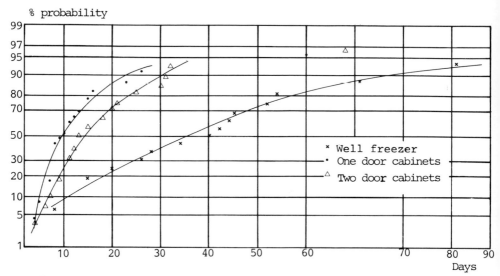

Fig. 111. Average storage time in home freezer cabinets in Australia, according to Sharp and Irving (1976).

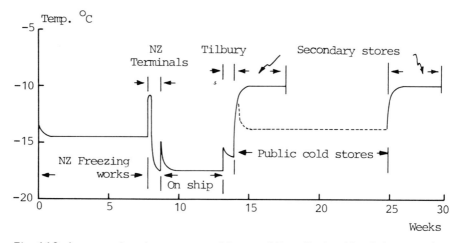

Fig. 112. Average time/temperature history of New Zealand lamb from works store to retail outlet in UK. After MIRINZ (1981).

International trade

Most time-temperature surveys of the freezer chain have been limited to national situations because of the obvious difficulties and extra expenditure involved in investigating the fate of exported products. Yet, the Meat Industry Research Institute of New Zealand has determined a typical time-temperature

Table 96. Estimated time-temperature experience of frozen beef, exported from Denmark. Data from Sven Qvist (1981).

	General	
Producers store	14-90 days	–30°C
Transport	1-2 days	–20°C
Freezer container	14-35 days	–22°C
	Saudi Arabia	
Freezer storage	30-60 days	–10 to –25°C
Refrigerated truck	1-2 days	–10 to –18°C
Sales cabinet	7-14 days	– 8 to –12°C
	Egypt	
Non-refrigerated truck	0.3 day	ambient temp.
Non-refrigerated sales facility	0.2 day	ambient temp.

Note: Meat is in short supply in Egypt and will normally be consumed the same day it is shipped from the freezer storage warehouse. In Saudi Arabia freezer storage of meat is required to be below –10°C.

history of exported frozen lamb as referred to above and indicated in Fig. 112. An industry such as, for instance, the Danish meat industry faces special problems in determining the time-temperature conditions to which exported frozen meats may be exposed since a great number of countries are involved. Table 96 suggests possible experiences of frozen meats shipped from Denmark to Egypt and Saudi Arabia as estimated by Sven Qvist, Danish Meat Products Laboratory (1981).

Temperature fluctuations

Since the early days of frozen foods, the technical literature has abounded with statements to the effect that it is very important to maintain freezer storage rooms at constant temperature, i.e. to avoid temperature fluctuations. These warnings seem to have been based on the general knowledge, to which reference was made above, that fluctuating temperatures under some circumstances may cause recrystallization and thus cause a change in some crystal systems. It was assumed that the small crystals, which were assumed to exist due to quick freezing, would grow into larger crystals with presumed adverse effects on product quality, cf. Fig. 13.

This warning is probably by and large exaggerated or even unnecessary. As mentioned above, most commercially-used freezer storage rooms maintain approximately constant temperatures, because temperature is comparatively easily regulated in large rooms. In addition, the frozen product is generally

Table 97. Fluctuating temperatures' influences on organoleptic scores for peas, okra and strawberries after 6 months. After Moleeratanond *et al.* (1979). (Score: 3-3.9 good quality; 2-2.9 second grade).

	Place of package in pallet	
	Internal	Corner
−23°C constant	3.75	3.37
−23°C fluctuating to −18°C	3.39	3.34
−21°C fluctuating to −18°C	3.37	3.33
−18°C fluctuating to −15°C	3.38	3.30

placed in large stacks, stacked on pallets or similar. This means that a very considerable frozen mass is involved, and that even comparatively large air temperature changes in the storage room would hardly affect product temperature. The main problem may exist near the doors of less well managed cold store facilities.

Guadagni and Nimmo (1958) showed that the effect of fluctuating temperatures on frozen strawberries and raspberries was the same as that obtained by the equivalent constant temperatures.

Tressler and Evers (1957) quote many instances where fluctuating temperatures *per se* had no other effect than that which could be calculated by adding the shelf life loss at each temperature for the total dwell period at that temperature.

Moleeratanond *et al.* (1979) carried out some experiments about this factor and found little quality change as a result of temperature fluctuation in freezer storage rooms, cf. Table 97.

They found in some cases a somewhat better quality in packages in the center of a pallet than in corners but this difference was independent of storage conditions. Similarly, Moleeratanond *et al.* (1981) demonstrated that considerable temperature fluctuations in freezer stores permitted to achieve energy savings - or due to poor temperature maintenance - had no discernible effect on the the quality and nutritive value of frozen boxed beef when it was stored for 12 months. It appears as if this subject area is in need of further investigations.

Problems of temperature fluctuations were discussed also under the consideration of additivity of the effect of various episodes of the freezer chain. One may also quote Hustrulid, Winther and Noble (1949), who showed that packaged, ground beef or ground pork were not markedly affected in quality by temperatures fluctuating from 0 to -10°F (-18 to -23°C) with 3 cycles per week for 6 months, compared to storage at a constant temperature of 0°F (-18°C). Only when unsatisfactory packaging material was used, the effect of temperature fluctuation on quality was quite marked due to surface desic-

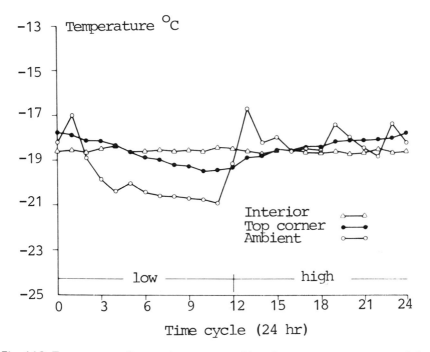

Fig. 113. Temperature fluctuation measured in a freezer storage room and the corresponding temperature in packages on a pallet of packaged strawberries in the room. Thermostat set at a low temperature the first 12 hours, in a higher position the next 12 hours. After Ashby *et al.* (1979).

cation. Since most frozen meat products in consumer-size packages are well packaged, it appears that the strong emphasis generally put on keeping the storage temperatures constant with very small fluctuation is quite unnecessary for this type of freezer goods.

Ashby *et al.* (1979) found very little change in product temperature even when temperature changes in the freezer store were considerable, see Fig. 113. Where products are unpackaged, e.g. raw material for later use, the situation is quite different. Here, fluctuations may cause surface desiccation, weight loss and eventually freezer burn.

Ulrich (1981) gives a very detailed review of data regarding the influence of temperature fluctuations on the quality of frozen food. In general he concludes that apart from a few special cases, storage at fluctuating temperatures has much the same effect as storage at a constant temperature, the latter being the "effective average" temperature of maximum and minimum temperatures during fluctuations.

Table 98. Amount of in-package ice crystals in weight per cent. After Van den Berg (1966).

In-package ice crystal	Number of packages		
	Peas	Other vegetables	All vegetables
0-2%	9	4	13
2-4%	10	8	18
4-6%	15	5	20
6-8%	6	0	6
8-10%	2	2	4
10-12%	0	1	1
12-14%	0	0	0
14-16%	1	1	2
16-18%	1	0	1
18-20%	1	0	1

While temperature fluctuations during wholesale storage are unlikely to represent any problem, more attention should probably be paid to what happens during the period of retail sales display. Here, temperature fluctuations are often large and frequent, and include such temperature zones where quality loss may be suffered much more quickly. One consequence of fluctuating temperatures is referred to in Table 91, in-package desiccation, which can be quite a serious problem at retail level. Van den Berg (1966) gives the data in Table 98 for random retail purchases in Ottawa in 1964.

It is seen that it is not uncommon that up to 8% of product weight evaporates and sublimates in the form of ice on the inside of a package or among loosely frozen products such as peas. This process takes place during freezing when the surface of the package is colder than its contents, and also any time during storage, especially at warmer temperatures, e.g. in freezer storage cabinets, when the surface is colder than the contents. The process is, of course, not reversible so repeated fluctuations result in a gradual accumulation of ice inside the package. This means that the product dries up and may even be subject to freezer burn. Consumers react quite strongly to this phenomenon and resist buying products with much visible frost inside the package. Further, since in-package ice is discarded in net weight control, serious problems with adherence to weight and measures regulations may occur. Also, according to Van den Berg (1966), eating quality is affected in that odour and flavour change.

The problem is often most noticeable in water-vapor impermeable packages. However, this is due to the fact that in other packages the water may evaporate more or less completely from the package and not be visible. Thus, it may cause less consumer reaction although the effect on eating quality and net weight will be the same or even greater.

The problem affects products consisting of small units more than those where each unit is larger, e.g. in-package desiccation and drying during freezer storage affects red currants more than cherries.

Products frozen in small pieces, e.g. peas, goulash meat, often receive an improved consumer acceptance if packaged in vacuum, covered by a gravy, dressing or the like, (Dalhoff and Jul, 1965, Philippon, 1981). One of the reasons for this is undoubtedly that such measures prevent in-package desiccation.

The smaller influence of temperature fluctuations in a bulk pack compared to a retail package sometimes recorded for fruits may be due to less in-package desiccation, as noted by Guadagni (1969).

Reference should be made also to investigations regarding the retention of vitamins of the B-group during fluctuating freezer storage temperatures. As mentioned earlier, no specific effect of temperature fluctuations was found.

The importance of temperature fluctuations and handling abuses in the retail stores are clearly an area in need of further clarifications.

Temperature requirements

Delaunay and Rosset (1981) indicate how, in France, attempts are made to differentiate for various products, according to their sensitivity, as regards the temperatures prescribed for storage and transport, cf. also Table 80.

Ulrich (1981) mentions that it may well be possible to obtain satisfactory quality of frozen foods at temperatures warmer than -18°C provided freezer storage temperature is not raised too much and as long as storage times are not excessive.

In this context, reference is made also to Fig. 112 and the fact that in so far as frozen meats are concerned, Australia and New Zealand have long accepted temperatures in the freezer chain considerably warmer than -18 to -20°C, i.e. often -10 to -15°C, as has also been the case in France and the Federal Republic of Germany in so far as products designated "congelé" or "gefroren" are concerned. Thus, as described by Delaunay and Rosset (1981), the concept "surgelé" was introduced in France at the same time as "quick freezing" in the USA. In France it applies to products which have been maintained at -18°C or colder at all times, and is thus contrasted with the "congelé" concept. It is generally assumed that products designated "surgelé" are of better quality since they have been stored at a colder storage temperature. However, this may often be due to better care throughout the manufacturing process for such products and not to temperature management. A similar distinction is made in the Federal Republic of Germany between products designated "gefroren", generally maintained at -12°C, and those designated "tiefgefroren".

As will be seen from the above, this distinction may be less meaningful for some products than it at one time was assumed to be. Products with reverse stability behave in a manner which even make the distinction meaningless or unfortunate; further, the time-factor and the PPP-factors have such over-riding influence on quality that characterizing a product according to storage temperature only may have very limited value as regards the product's actual quality as experienced by the consumer. Also, as Fig. 29 shows, a product like frozen chicken is so relatively stable at warm freezer temperatures that there is hardly any need to store it at very cold temperatures. This is probably the reason why in France most poultry is sold "surgelé" not "congelé", cf. Tables 1 and 80.

In this context, it may be mentioned that in the UK, a similar distinction is made between the terms "frozen" and "deep frozen", although no legal definition seems to exist. In English, "deep frozen" is often synonymous with the German terms "schnell gefroren" and "tiefgefroren" or even "tiefge-kühlt".

Temperature recommendations

In the earlier days, there was a tendency to recommend the lowest possible temperature for frozen foods, e.g. -30°C at producer and wholesale level, and -18°C for retail stores and home freezers. This has been somewhat modified in recent years because of energy considerations. When one views the widely different TTT-curves, which various types of produce have due to PPP-factors, it becomes evident that the choice of temperature must of necessity be a compromise. In a few cases, one may make product-specific recommend-ations, e.g. possibly -40°C for non-blanched fruits and vegetables for further processing in the same plant, -28°C for fish at the producer and wholesale level, -15°C for frozen chicken at wholesale level, and -8°C for storage of unpacked Wiltshire bacon, etc. For the general frozen food industry, few means exist for product-specific considerations. Here one may recommend -22°C for the wholesale level, reasonable precautions against excessive temperature rise during transport, and retail cabinets operating at no warmer than -10°C in any one point below the load line. Finally, one must hope that more meaningful advice may be made available to the industry, trade and consumers in order that they may have some notion about very temperature sensitive products, e.g. fruits in sugar, and very time sensitive ones, i.e. cured meats.

Actual Shelf Life Calculations

Data for the freezer chain

For the purpose of illustration, the author considered it relevant to carry out some calculations of shelf life losses in what may be assumed to be the contemporary freezer chain. For this the author collected data on times and temperatures in various periods of frozen food distribution. As already mentioned, actually determining such data is cumbersome and time consuming, and only a limited number of such studies are available. On the basis of data by Lorentzen (1971), Middlehurst and Richardson (1972), Boegh-Soerensen and Bramsnaes (1977), Olsson and Bengtsson (1972), Sharp and Irving (1976), and data made available by Jean Philippon, France, the author estimated times and temperatures in a hypothetical freezer chain, which may be somewhat typical of the time and temperature experience to which frozen products are exposed in national distribution in North America and Western European Countries. The data are illustrative only. They have been somewhat arbitrarily put together from data collected in different countries for different periods and times and using somewhat different criteria. However, they may be sufficiently typical to illustrate the present situation. As exemplified in Table 96 and Fig. 112, conditions are vastly different in some export situations.

For the purpose of such calculations, the author chose to base his estimates on the time and temperature condition data given in Figs 114 to 123.

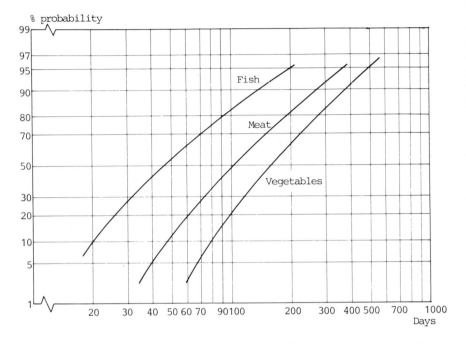

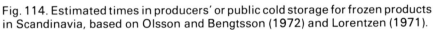

Fig. 114. Estimated times in producers' or public cold storage for frozen products in Scandinavia, based on Olsson and Bengtsson (1972) and Lorentzen (1971).

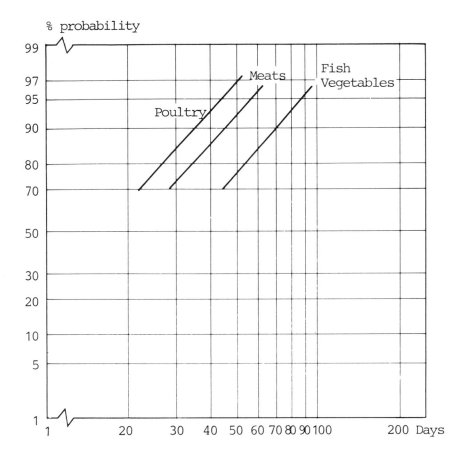

Fig. 115. Probability of storage time at wholesale level in Sweden. Based on Olsson and Bengtsson (1972).

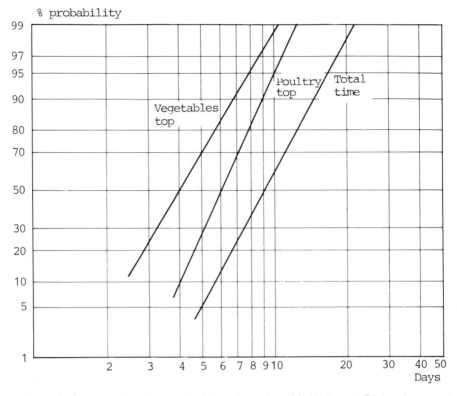

Fig. 116. Storage time in retail cabinet, based on Middlehurst, Richardson and Edwards (1972).

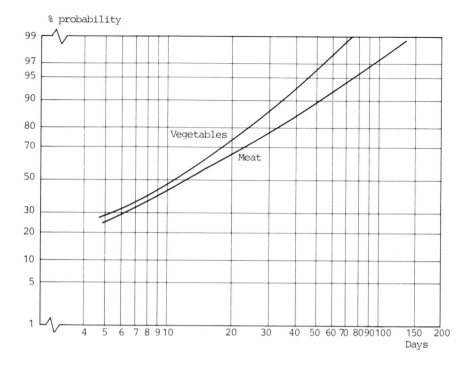

Fig. 117. Storage time in home freezers, based on Sharp and Irving (1976).

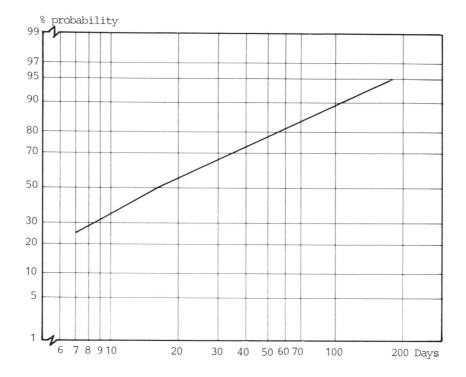

Fig. 118. Probability of products' storage time in home freezer, based on Redstrom (1971).

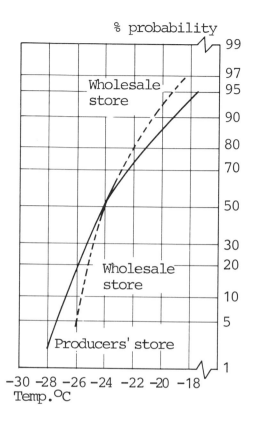

Fig. 119. Temperature distribution in producers' store in Norway in 1970 (Lorentzen, 1971), and in wholesalers' cold stores, based on Spiess *et al.*, (1977).

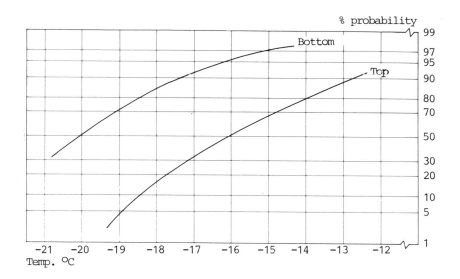

Fig. 120. Temperature distribution in frozen poultry in freezer cabinets in France, based on Comité Interministériel de l'Agriculture et de l'Alimentation (1980).

% probability

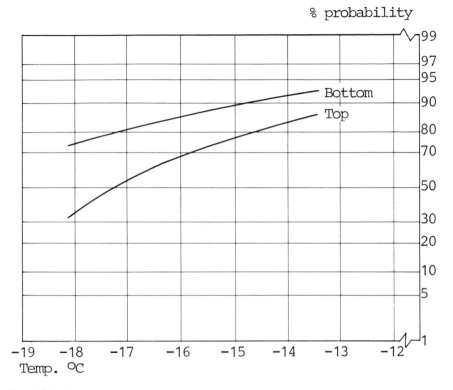

Fig. 121. Temperature distribtution in packages with haricot verts, beans, in freezer cabinets in France, based on Comité Interministériel de l'Agriculture et de l'Alimentation (1980).

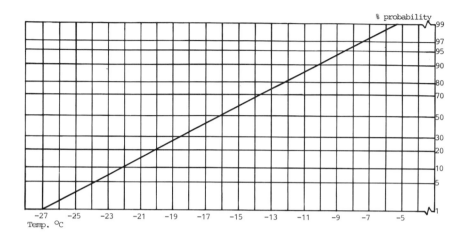

Fig. 122. Temperature distribution in home freezers, based on Sharp and Irving (1976).

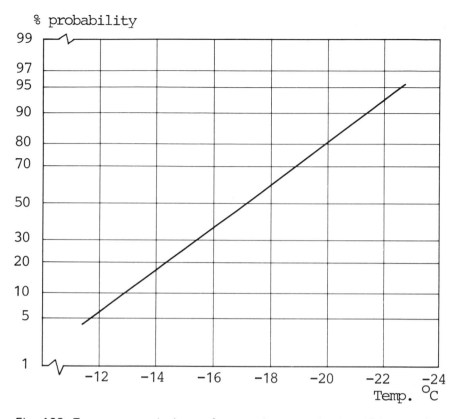

Fig. 123. Temperatures in home freezers in towns in the USA, based on Redstrom (1971).

Probability of exposure

In the above, the data for time and temperatures in the freezer chain are given as a distribution curve showing the probability of a sample being exposed to a certain temperature or time condition. To enable meaningful calculations to be made, it is necessary to decide, again somewhat arbitrarily, what degree of "risk" one should use in calculations of probable end product quality. As mentioned above, the author (Dalhoff and Jul, 1965) in calculations presented in 1963 used such conditions where no more than 5% of all samples would be exposed to more adverse conditions. It was conceded that this was a rather stringent type of calculation, since the probability of one and the same package being exposed to the worst temperature and time conditions at both producer, wholesale, and retail level is indeed very small.

Gustav Lorentzen, Norway, cf. also Fig. 75, has in his calculations used the 90% probability conditions. Some have calculated the conditions to which the average product will be exposed. The latter is probably too optimistic an approach. Leif Boegh-Soerensen, Denmark, has suggested that in such calculations one should use such data where only 25% of all samples will be exposed to more adverse conditions; the author has followed this suggestion in the calculations cited below.

TTT-data

For calculations, data on time-temperature tolerance (TTT) of various foods were also required. Since the very extensive work carried out by the USDA Western Regional Research Center in Albany (now Berkeley), and the general interest in determining the TTT of various foods in the years following that project, comparatively few data have been published on this subject. Well defined recent data have been made available by Boegh-Soerensen (1975) on broilers and by Flemming Lindeloev (1978) on pork, untreated or cured. For fish, fruits and vegetables only older data seem available. Therefore, such older data have had to be used in the present context.

Actual estimates of end product quality

On the basis of the TTT-data collected for various products and reproduced in Figs 124, 126, 128, 130, 132, 134, 136, 138, 140, 142, 144, and using the time-temperature probability conditions shown in Figs 114 - 123, the author calculated the estimated loss of shelf life for a number of products. The results are given in the calculations and diagrams immediately following the time-temperature curve for each product. It should be noted that in the diagrams

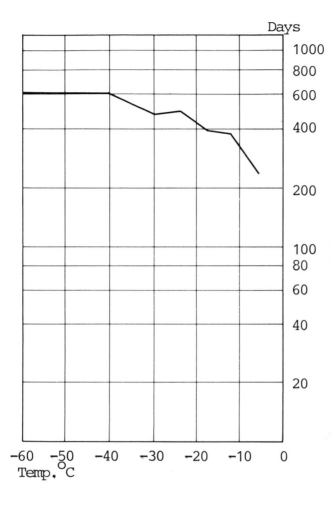

Fig. 124. Acceptability time for vacuum packaged, sliced, untreated pork. After Lindeloev (1978), cf. Fig. 25.

showing shelf life loss, various scales have been used. Since the area under the curve indicates shelf life loss, it is important to keep this in mind. As will be seen, in the time-temperature exposure assumption, a period of storage in home freezers was included.

Limitations of data

In considering this series of data, it must be kept in mind that product,

Table 99 Losses of shelf life of vacuum packaged sliced untreated pork.

Stage	Time	Temp.	Acceptability	Loss per day	Loss
Producer	170 days	−22°C	470 days	0.00213	0.36
Transport	2 days	−14°C	370 days	0.00270	0.01
Wholesale	32 days	−21°C	410 days	0.00244	0.08
Transport	1 day	−12°C	280 days	0.00357	0.00
Retail	9 days	−14°C	370 days	0.00270	0.02
Transport	0.1 day	−3°C	200 days	0.00500	0.00
Subtotal	214 days				0.47
Home freezer	26 days	−13°C	300 days	0.00333	0.09
Total	240 days				0.56

processes and especially packaging have been much improved over the later years. Therefore, older data, such as those used for fish, fruits and vegetables, are to be taken with some considerable reservations.

Another factor should be mentioned. All data available have been collected in laboratory experiments. This means that the abuse, especially the careless or incorrect handling, which is experienced in parts of the freezer chain e.g. during loading and unloading, in retail stores and in consumers' handling is not included in the tests. To take some examples, packaging may be damaged during transport and handling or when the product is placed in the retail cabinet, or the period in the retail cabinet may involve considerable temperature fluctuations. These have not been included in experiments. One

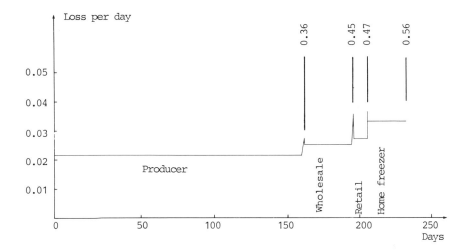

Fig. 125. Shelf life loss of vacuum packaged, sliced, untreated pork.

of the aims of future research must be to get some more data regarding these factors and to modify test conditions accordingly.

Mindful of these reservations, the author has used these data for some storage life loss calculations.

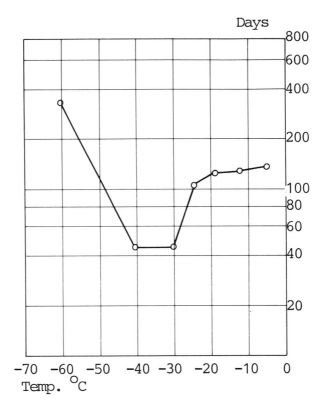

Fig. 126. Acceptability time for sliced, smoked bacon, packaged in oxygen permeable film (Polyethylene). After Lindeloev (1978).

Table 100. Loss of shelf life of sliced, smoked bacon in polyethylene packages.

Stage	Time	Temp.	Acceptability	Loss per day	Loss
Producer	180 days	–26°C	85 days	0.01176	2.1
Transport	2 days	–14°C	128 days	0.00781	0.0
Wholesale	32 days	–25°C	100 days	0.01000	0.3
Transport	1 day	–12°C	130 days	0.00769	0.0
Retail cabinet	12 days	–15°C	126 days	0.00793	0.1
Transport	0.1 day	–3°C	140 days	0.00714	0.0
Subtotal	227 days				2.5
Home freezer	17 days	–12°C	130 days	0.00769	0.1
Total	244 days				2.6

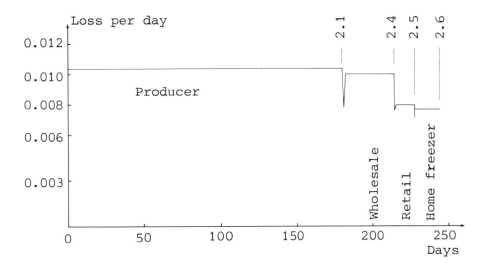

Fig. 127. Shelf life loss of sliced smoked bacon in permeable film.

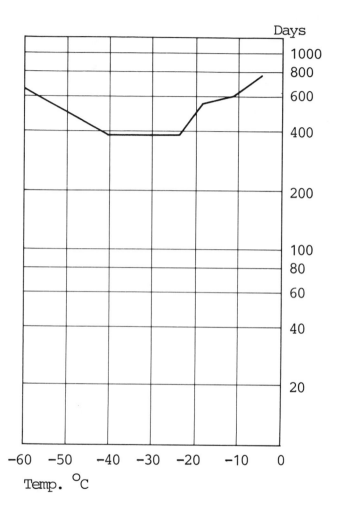

Fig. 128. Acceptability time for vacuum packaged, smoked, sliced bacon, 7% salt-in-brine. After Lindeleov (1978).

Table 101. Shelf life loss for smoked vacuum packaged bacon, 7% salt-in-brine.

Stage	Time	Temp.	Acceptability	Loss per day	Loss
Producer	180 days	−26°C	360 days	0.00278	0.50
Transport	2 days	−22°C	450 days	0.00222	0.00
Wholesale	32 days	−25°C	380 days	0.00263	0.08
Transport	1 day	−12°C	480 days	0.00208	0.00
Retail cabinet	12 days	−21°C	440 days	0.00227	0.03
Transport	0.1 day	−3°C	550 days	0.00182	0.00
Subtotal	227 days				0.61
Home freezer	12 days	−12°C	480 days	0.00208	0.02
Total	239 days				0.63

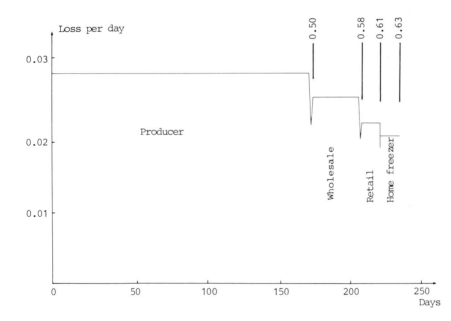

Fig. 129. Shelf life loss of sliced, vacuum packaged bacon, 7% salt-in-brine.

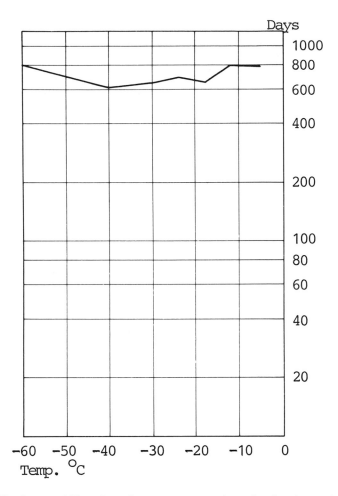

Fig. 130. Acceptability time for vacuum packaged, sliced, smoked bacon (Lindeloev, 1978), cf. Fig. 39.

Table 102. Shelf life loss for vacuum packaged, sliced, smoked bacon.

Stage	Time	Temp.	Acceptability	Loss per day	Loss
Producer	170 days	–26°C	700 days	0.00143	0.24
Transport	2 days	–22°C	650 days	0.00154	0.00
Wholesale	32 days	–25°C	680 days	0.00147	0.05
Transport	1 day	–12°C	800 days	0.00125	0.00
Sales cabinet	9 days	–21°C	630 days	0.00159	0.01
Transport	0.1 day	–3°C	750 days	0.00133	0.00
Subtotal	214 days				0.30
Home freezer	26 days	–20°C	650 days	0.00154	0.04
Total	240 days				0.34

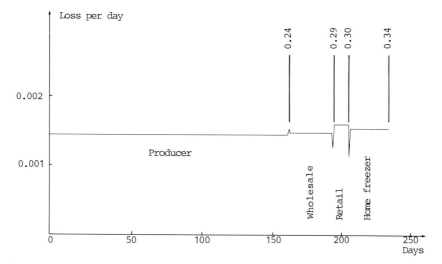

Fig. 131 Shelf life loss of vacuum packaged, sliced, smoked bacon. There is a quite significant difference in loss of shelf life compared with the product illustrated in Figs 128 and 129, even though the acceptability diagrams appear to be quite similar.

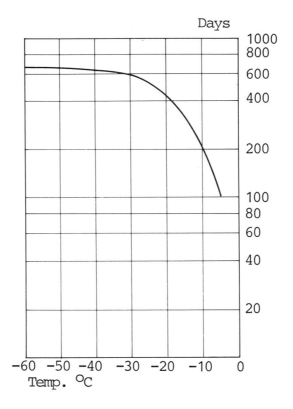

Fig. 132. Acceptability time for meat balls packaged in oxygen permeable film
After Lindeloev (1978).

Table 103. Loss of shelf life of meat balls in polyethylene film.

Stage	Time	Temp.	Acceptability	Loss per day	Loss
Producer	170 days	–22°C	450 days	0.00222	0.38
Transport	2 days	–14°C	310 days	0.00323	0.01
Wholesale	32 days	–21°C	400 days	0.00250	0.08
Transport	1 day	–12°C	220 days	0.00455	0.00
Retail	9 days	–14°C	230 days	0.00435	0.04
Transport	0.1 day	–3°C	70 days	0.01429	0.00
Subtotal	214 days				0.51
Home Freezer	26 days	–13°C	210 days	0.00476	0.12
Total	240 days				0.63

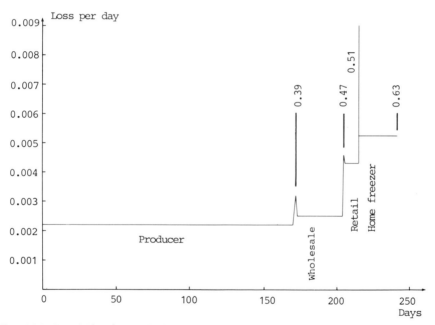

Fig. 133. Shelf life of meat balls packaged in polyethylene.

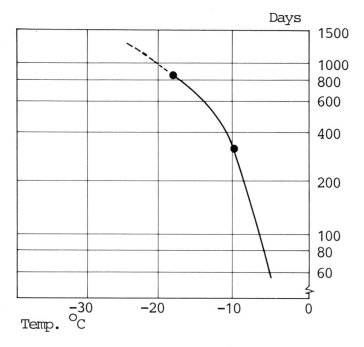

Fig. 134. Acceptability time for whole broiler chickens. After Boegh-Soerensen (1975).

Table 104. Loss of shelf life of chicken broilers in polyethylene.

Stage	Time	Temp.	Acceptability	Loss per day	Loss
Producer	170 days	–23°C	1300 days	0.00077	0.13
Transport	2 days	–14°C	600 days	0.00167	0.00
Wholesale	30 days	–22°C	1200 days	0.00083	0.02
Transport	1 day	–12°C	480 days	0.00208	0.00
Retail	7 days	–16°C	720 days	0.00139	0.01
Transport	0.1 day	–4°C	30 days	0.03333	0.00
Subtotal	210 days				0.16
Home freezer	27 days	–13°C	550 days	0.00182	0.05
Total	237 days				0.21

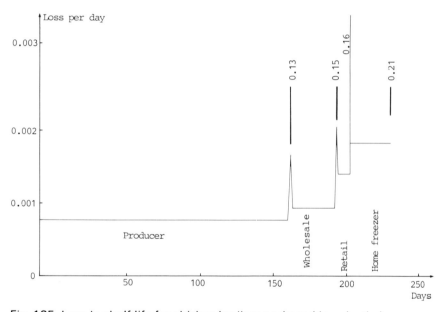

Fig. 135. Loss in shelf life for chicken broilers packaged in polyethylene.

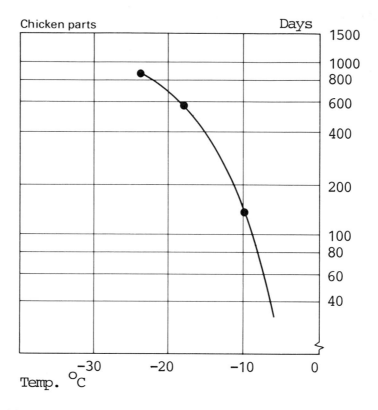

Fig. 136. Acceptability time for chicken parts (thighs), packed in foodtainer with polyethylene film. After Boegh-Soerensen (1975).

Table 105. Loss of shelf life of chicken parts packed in polyethylene.

Stage	Time	Temp.	Acceptability	Loss per day	Loss
Producer	170 days	–23°C	850 days	0.00118	0.20
Transport	2 days	–14°C	350 days	0.00286	0.00
Wholesale	30 days	–22°C	800 days	0.00125	0.04
Transport	1 day	–12°C	250 days	0.00400	0.00
Retail	7 days	–16°C	400 days	0.00250	0.02
Transport	0.1 day	–4°C	20 days	0.05000	0.01
Subtotal	210 days				0.27
Home freezer	27 days	–13°C	300 days	0.00333	0.09
Total	237 days				0.36

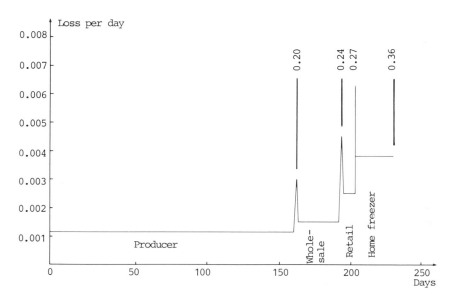

Fig. 137. Loss of storage life for chicken parts packaged in polyethylene.

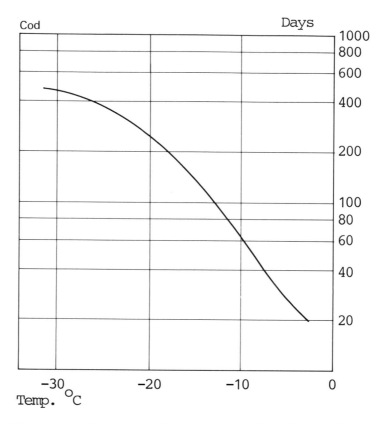

Fig. 138. Acceptability time for frozen cod, based on Gutschmidt, quoted by Spiess (1980), cf. also Fig. 53.

Table 106. Loss of shelf life of frozen cod.

Stage	Time	Temp.	Acceptability	Loss per day	Loss
Producer	90 days	−22°C	300 days	0.003333	0.03
Transport	2 days	−14°C	125 days	0.00800	0.02
Wholesale	50 days	−23°C	310 days	0.00322	0.16
Transport	1 day	−12°C	90 days	0.01111	0.01
Retail	13 days	−11°C	77 days	0.01299	0.17
Transport	0.1 day	−3°C	18 days	0.05555	0.00
Subtotal	156 days				0.39
Home freezer	20 days	−13°C	100 days	0.01000	0.20
Total	176 days				0.59

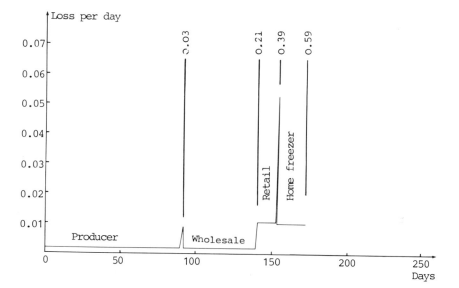

Fig. 139. Loss of shelf life of frozen cod.

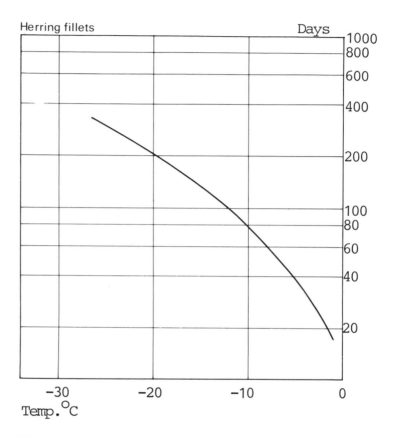

Fig. 140. Acceptability time for frozen herring fillets wrapped in polyethylene in impregnated cartons, based on Löndahl (1973).

Table 107. Loss of shelf life of frozen herring fillets, polyethylene wrapped, in impregnated cartons.

Stage	Time	Temp.	Acceptability	Loss per day	Loss
Producer	90 days	–22°C	250 days	0.00400	0.36
Transport	2 days	–14°C	120 days	0.00833	0.02
Wholesale	50 days	–23°C	260 days	0.00385	0.19
Transport	1 day	–12°C	98 days	0.01020	0.01
Retail	13 days	–11°C	86 days	0.01163	0.15
Transport	0.1 days	–3°C	25 days	0.04000	0.00
Subtotal	156 days				0.73
Home Freezer	20 days	–13°C	110 days	0.00909	0.18
Total	176 days				0.89

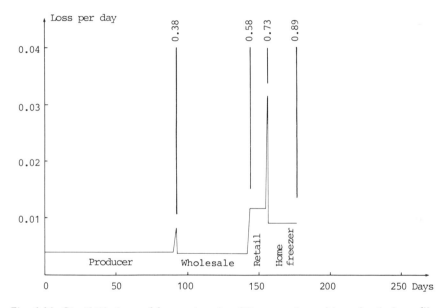

Fig. 141. Shelf life loss of frozen herring fillets, packaged in polyethylene film and placed in impregnated cartons.

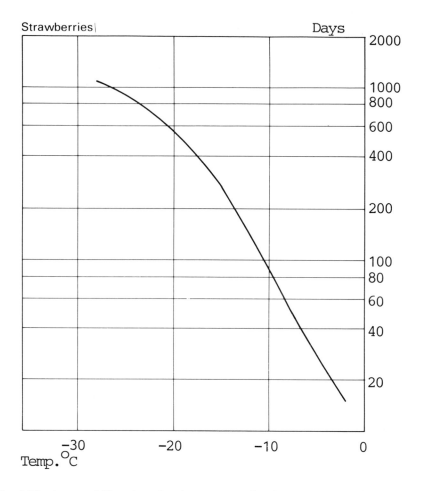

Fig. 142. Acceptability time for frozen strawberries, based on Gutschmidt, quoted by Spiess (1980).

Table 108. Loss of shelf life of frozen strawberries.

Stage	Time	Temp.	Acceptability	Loss per day	Loss
Producer	250 days	−22°C	660 days	0.001515	0.38
Transport	2 days	−14°C	220 days	0.004545	0.01
Wholesale	50 days	−23°C	710 days	0.001408	0.07
Transport	1 day	−12°C	140 days	0.007143	0.00
Retail	21 days	−11°C	110 days	0.009091	0.19
Transport	0.1 day	−3°C	18 days	0.055555	0.01
Subtotal	324 days				0.66
Home freezer	20 days	−13°C	180 days	0.005555	0.11
Total	344 days				0.77

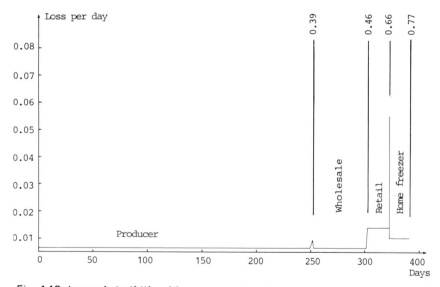

Fig. 143. Loss of shelf life of frozen strawberries.

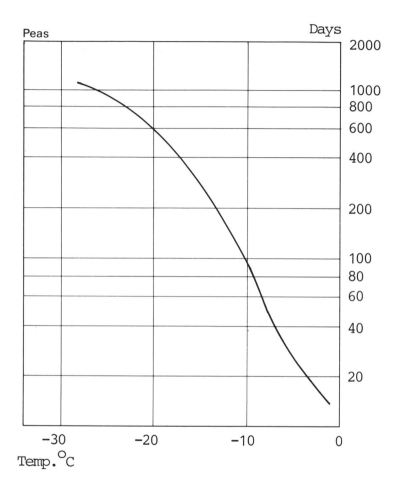

Fig. 144. Acceptability time for frozen peas, based on Gutschmidt, quoted by Spiess (1980).

Table 109. Loss of shelf time of frozen peas.

Stage	Time	Temp.	Acceptability	Loss per day	Loss
Producer	250 days	−22°C	700 days	0.001429	0.34
Transport	2 days	−14°C	230 days	0.004348	0.00
Wholesale	50 days	−23°C	730 days	0.001370	0.06
Transport	1 day	−12°C	145 days	0.006667	0.00
Retail	21 days	−11°C	115 days	0.008696	0.18
Transport	0.1 day	−3°C	19 days	0.052632	0.01
Subtotal	324 days				0.59
Home freezer	20 days	−13°C	204 days	0.05405	0.11
Total	344 days				0.70

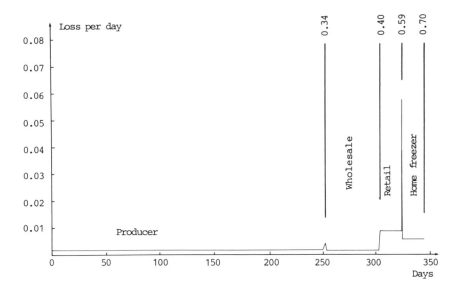

Fig. 145. Shelf life loss for frozen peas.

Conclusions of calculations

A review of these data shows as an obvious conclusion that sliced bacon in a permeable film is not suited for freezing. The other meat products seem to have shelf life losses of 0.3 to 0.6, cf. Figs 125, 129, 131, and 133; it is concluded that these products, packaged and treated as in the tests on which the time-temperature-tolerance data were based, would be perceived by the consumer as fully acceptable.

For the poultry products, both whole broilers and broiler parts, Fig. 135 and Fig. 137 only 0.2 to 0.4 of shelf life is lost by the exposure to a frozen food distribution as here assumed. This suggests - as is general experience - that these products keep very well during normal frozen storage, although the tests on which these time-temperature-tolerance data were based, were carried out on chicken packaged in rather permeable film. These calculations confirm that it is understandable that these products are normally packaged only in such material. It has often been recommended that they be packed in a better, oxygen-impermeable film. This does, indeed, result in a better shelf life as indicated in Figs 29 and 30, but the above calculations suggest that such precautions are not really necessary. It may be noted that the situation is different for turkey products; these are often packed in cry-o-vac or other films with very low oxygen-permeability. The explanation is likely to be that rancidity is a real probability in frozen turkey products. In practice, it is probably this superior durability of frozen chicken which has resulted in these products mostly being traded as "congelé" or "gefroren" in France and the Federal Republic of Germany respectively.

For fish, fruit and vegetables, i.e. Figs 139, 141 and 143, 0.6 to 0.9 of shelf life seems to be lost during the passage of the freezer chain. This seems to be rather high. One explanation may be that the time-temperature-tolerance data, on which these calcuations are based, are old data. Recent improvements in products, processes, and packaging were not taken into account.

As already mentioned, these calculations must be taken with the above-mentioned reservation. The time-temperature-tolerance data were obtained by laboratory experiments and did not include the damage to the packaging material which may occur during handling, nor did they take into account the temperature fluctuations which are likely to occur in the retail cabinet. As indicated in Fig. 104, this is a factor on which more data are needed.

Data from a large survey, referred to above and reproduced in Table 86, carried out as a collaborative experiment in support of Codex Alimentarius discussions, showed that only 2 to 11% of all the products that has passed through the regular freezer chain were not acceptable to presumably rather discerning laboratory panels. Considering this, the results quoted above seem to show that the method used for calculation here and especially the choice of the 0.75 probability level, may be a useful tool in considerations regarding the

Keep this spot yellow

Fig. 146. A suggested marking of a frozen food package with a temperature sensitive material. The spot may change colour from yellow to red when temperature gets warmer than -16°C. After Blixt (1983).

shelf life of frozen foods, cf. the Time-temperature surveys section above.

However, it is quite obvious that further research in this area is indicated. Thus, it would be useful to carry out acceptability tests on known products after they have passed through the regular freezer chain, i.e. which are purchased in stores, from products where the time-temperature-tolerance and the initial quality is known. If this were done, one would be able to determine whether the calculation method should be modified. Until such data become available, the author recommends that the probability factor of 0.75 be used. Further data on time temperatures in the freezer chain and more up-to-date TTT-diagrams for various products would be useful.

Temperature indicators

Blixt (1983) indicates that the Swedish I-Point corporation is considering printing a spot on packages for frozen food which, by its colour, would indicate if the package is below a predetermined temperature, cf. Fig. 146. Such a system might be of considerable help to frozen food handlers.

Clark (1983) indicates that the Hallcrest Products Corp., Glenview, IL, USA, is developing a system of cholesteric liquid crystals which may be incorporated into coatings which can be placed on loads of frozen foods. They change colour reversibly and rapidly, e.g. may be black at -15°C, red at -14°C, green at -13°C, and blue at -12°C. They are available for any temperature from -40°C to 250°C. One idea is to use them as warning labels, e.g. a black sticker on a pallet of frozen foods may suddenly read in red, green or blue: "Place me colder!"

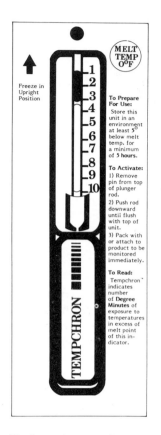

Fig. 147. Device which will show degree of temperature-abuse over a pre-determined temperature. A reading of say 10 indicates exposure to a temperature 10°F over the critical point for 1 minute or 5°F over the same for 2 minutes, etc. (Andover Laboratories, 1983).

Temperature abuse indicators

For guidance on whether a frozen product has been exposed to temperature abuse, a small ball or cube of ice has often been placed in the frozen food pack. If it disappeared or at least lost its shape, one would know that the product had been above 0°C. According to the Grocer (1982), a South African insurance company has used a coloured marble-size ball in a plastic bag for a similar purpose. If some chemical is added to the water before its being frozen, it may be used to detect exposure to temperatures colder than 0°C.

It may also be possible to use irreversible heat sensitive colours for this purpose, the colour changing mechanism being activated by freezing.

Blixt (1983) has described the use of inexpensive indicators, which change colour irreversibly if a predetermined temperature, e.g. the melting point of a product frozen with sugar is exceeded, cf. also the section: Improved cabinet management above. Their use would enable a receiver of frozen foods to verify whether a predetermined level of temperature has been exceeded during shipment, etc. They would normally be placed on pallets, shipping cartons or the like.

Temperature abuse integrators

Some work has been devoted to measuring the extent of temperature abuse, i.e. how much a predetermined temperature was exceeded. Thus, Fig. 147 shows a device which integrates the time and the severity of the abuse, i.e. the sum of the multiples of degrees in degrees Fahrenheit over a critical point and the time of excess in minutes. In the frozen food range, these devices are available for 0°F(+17.8°C), 10°F(-12.2°C), and 20°F(-6.7°C), but may be supplied with other melting points. The device called Tempchron, was developed by Andover Laboratories, Weymouth MA, USA. It functions simply by the incorporation of liquids with various melting points and diffusion rates.

Manske (1983) describes a system developed by the 3M Company, St. Paul, Minnesota, USA. It is called the Monitor-mark; it works in a somewhat similar manner. So far it seems to have been designed only for refrigerated, not for frozen foods; the melting material is fatty acids with a dye.

Time-temperature integrators

Much effort has been devoted to developing more sophisticated time-temperature integrators which could indicate the accumulated effect on shelf life for a certain food after its passage through the freezer chain. Thus, June Olley (1978) described an instrument which follows the time and temperature of a product, integrates the time-temperature effect and indicates the shelf life loss once the TTT-curve for the product has been determined. Such instruments may be useful for experimental purposes. Since they involve sensitive instruments, they are not suited for large scale tests under commercial conditions.

More recently, Storey and Owen (1982) have described simple and impact-resistant devices of the same type, so far developed for use at temperature around the freezing point, i.e. for chilled foods.

Efforts have been devoted to designing simple devices which actually might be placed in or on each package of frozen food and which would show the integrated effects of the time and temperature to which the package has been

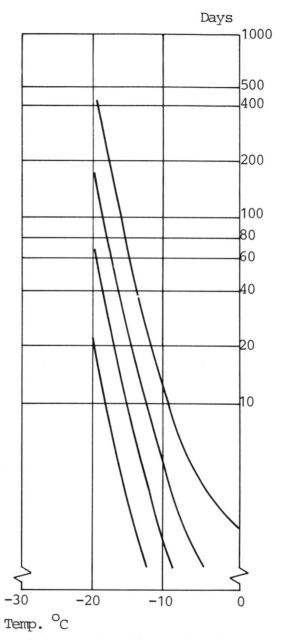

Fig. 148. Characteristic curves for 4 different I-Point integrators. The diagram shows for each type the time-temperature combination at which the colour in the indicator's window (see Fig. 149) corresponds to that of the area designated "uncertain". After K. Blixt (1983).

exposed. Again, it is only possible to devise such an indicator if the TTT-characteristics for the product are known. Some such devices are described by Renier and Morin (1962), i.e. an early development by the Honeywell Corporation. Others have been developed since.

Carseldine and Weste (1976) have described some low cost electronic devices which may be used to integrate a product's time-temperature history according to a variety of time-temperature functions, see also Plessey (1972). These devices require a simple instrument for reading-out the data that are accumulated and integrated in them.

Recent reviews of devices designed for such purposes were made by June Olley (1976), Byrne (1976), Kramer and Farquhar (1977 and 1982), Schubert (1977), and Mistry and Kosikowski (1983).

Even indicators such as the last ones mentioned are too expensive to be placed in each retail package of frozen food, so as to be of guidance for the retail trade and for the consumers. In order to achieve such utility, one would have to design a device so simple that it could be printed on each retail container of a frozen product. It would be interesting if such a system could be developed. It might act by a colour change or similar process and indicate approximately the severity of the time-temperature experience to which each package had been exposed. With the recent spectacular development in temperature sensitive colours it is not inconceivable that a system of that nature could be developed. One problem lies in that the indicator would have to be printed on the package, but the colour changing mechanism should be activated only when the package is frozen, e.g. by the process of ice crystal formation or the like. Some extra safeguards would be required, otherwise the integrator might react too rapidly to brief periods of exposure to heat, e.g. when a packaged is removed from a retail cabinet for brief inspection.

Some such development is already under way. Thus, Rose (1981) reviews the development of the above-mentioned Swedish time-temperature monitor based on an enzyme reaction. Farquhar (1981) discusses development in the use of both time-temperature recorders and integrators and gives further data for some such integrators shown in Fig. 149, containing an enzyme and a substrate, the latter with a pH indicator. The integrator is activated by breaking a seal between the two capsules and mixing the ingredients.

Earlier versions of this device are described by CRIOC (1982) and recommended for use in frozen food shipments in a pamphlet distributed by the centre in Brussels for research and information on consumers' organisations.

Blixt (1983) gives further data on these integrators which have the form of a self-adhesive, embossed label, with a window showing the colour of the substrate. The colour change is gradual, cf. Fig. 149. These may be developed for many different TT-characteristics.

The concept behind the design of these various integrators is that the temperature characteristics of various products can be fitted fairly well into

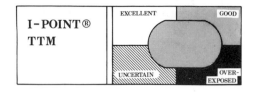

Fig. 149. I-Point time-temperature integrator. The four quarants are different shades of brownish-green, the time-temperature dependent colour appears in the window in the centre. After K. Blixt (1983).

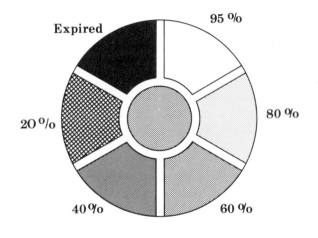

Fig. 150. Suggested time-temperature integrator for frozen foods. The outer segment matching the colour of the time-temperature sensitive dye in the center round spot will indicate remaining shelf life. After Fields and Prusik (1983).

various specific groups. The time-temperature data given above, e.g. Figs 124, 126, 128, 130, 132, 134, 136, 138, 140, 142 and 144, suggest that this is rather unlikely to be possible in all cases cf. Fig. 148.

Fields and Prusik (1983) describe another development in this field by Allied Corporation, Morristown, N.J., USA. It is based on colourless diacetylene monomers and similar materials which form highly coloured compounds with Arrhenius type kinetics. They may be adopted for many different time-temperature characteristics. They can be formulated into inks which may be printed on containers, cf. Fig. 150, where the degree of colour change would indicate remaining shelf life.

Where any such device is used on the outside of a pallet or in any similar way, it is important to keep in mind that the surface temperature reacts much quicker to changes in environmental temperature than that of the bulk of the product. One may have to bear this factor in mind when selecting integrators. It may be that a different time-temperature characteristic of the integrator than that of the product will have to be used. Otherwise, many readings may be made which are not indicative of what happened to the bulk of the product.

Sanderson-Walker (1979b) discusses the use of such integrators. Unless a breakthrough in design as suggested above occurs, he considers them too expensive for retail packages but useful for test shipments, on large pallets or other large units. He also calls attention to a possible undesirable side effect which their introduction might have. If integrators on retail packages showed a gradual change, which could be interpreted by the consumer, the latter could be expected to always select the products with most storage life intact, *viz.* this might result in what he calls a "last-in-first-out" effect. However, many countries at present require open date labelling of frozen foods. This could have been expected to have the same effect but in practice has not.

Time-temperature recording

For experimental and control purposes it might be useful if very small devices could be developed which could record time and temperature. With such devices, it would be comparatively easy to determine the time and temperature characteristics of the various components of the freezer chain and thereby determine not only the probable quality of the product, which can be determined organoleptically anyway, but to determine where in the freezer chain the greatest shelf life loss has taken place and thus, where corrective measures should be attempted.

Mills (1981) describes a time-temperature recording device, made available by Ryan Instruments Inc., Kirkland, WA., USA, on a rental basis. It has up to 90 days running time and may be calibrated -25°C to 35°C.

One device which was available for a period was a low-priced, disposable time-temperature mechanical recording apparatus which may be activated by the shipper and placed with a shipment, e.g. in a frozen food container. At the final destination, a time-temperature chart was obtained for each shipment. The device is no longer being manufactured.

These mechanical instruments may soon be replaced by electronic devices. It appears as if it might be very easy with the aid of micro chips, a quartz clock and calendar and a temperature sensing device to arrange for recording and storing time and temperature data for subsequent print-out in an appropriate device. It should be no problem to make such devices smaller than the usual combined pocket calculators and clocks, and as inexpensive. It should be possible also to make them so light that their presence would not interfere markedly with heat transmission characteristics. If they were designed as a self-mailing, postage-paid-by-addressee, package and a price for their return were offered, a whole new insight into the conditions of the frozen food chain might soon be gained. What may be a forerunner of such devices is made by the ETM company in Stockholm, which makes devices measuring 110 x 70 x 60 mm, usable in the temperature range from -50 to 150°C (ETM, 1983).

J.D. de Graaff, the Netherlands, has described such an instrument. Only prototypes exist but production is expected to begin shortly by the Thermo Electric Company, Warmond, The Netherlands.

Date Labelling

Expiry date

It is common practice in the frozen food industry to mark frozen foods with an expiry date. It is always quite difficult for manufacturers to decide what expiry date to put on a package. Manufacturers may carry out TTT-determinations for their products, e.g. such as shown above. However, as seen by the above review of time and temperature surveys in the freezer chain, manufacturers have no means of knowing the conditions to which each package will be exposed with regard to storage temperature. Therefore, they must base their date labelling on their knowledge of the characteristics of each product and the general impression of the time-temperature conditions during storage and distribution to which that particular product is likely to be exposed, i.e. base their decisions on calculations such as the above, accepting a certain level of "risk", and, of course, on experience.

"Use before" marking

Some authorities require that frozen food products be marked with a "use before" or "use by"date. This seems to be an unfortunate choice. It generally leads to food inspectors finding that products with a date expired should be destroyed. In some countries, e.g. Saudi Arabia, such a procedure is mandatory. Even regardless of any liberal interpretation on the part of food inspectors, consumers will assume that a product where the "use by" date is expired, is seriously deteriorated in quality or even unwholesome. Figure 31 and the many additional TTT-diagrams above illustrate how quality change is very gradual. The distinction between stability time or high quality life and acceptability time or practical storage life is also somewhat arbitrary, and the limit for acceptability may be perceived differently by different products and for different groups of manufacturers, consumers or authorities. Thus, a very

clearcut expiry date is in reality meaningless and is likely to force manufacturers to be too conservative in their indication of keeping time of their
products, thereby depriving the consumers, the trade and themselves of one of
the very advantages which food freezing presents, namely providing foods of
known very long durability. They may also become too careless and give long
shelf life times since it is general experience that frozen foods are normally sold
before the expiry date with which they are marked. In this case, too, date
marking is of no real use for the consumer. In no case, of course, will the fact
that the date is expired lead to any unwholesomeness of the food.

"Best before" date

A better choice seems to be the practice which is now accepted in the EEC
Labelling Directive, which requires labelling with "best before" date. Even in
this case, however, it is difficult to explain to consumers the exact meaning of
the designation.

Packaging date

Many consumers and authorities in some countries insist that products should
be marked also with the date the product was packed. In practice this can
often seem somewhat meaningless, because, as indicated below, many food
products are frozen to be thawed and used again for frozen products, in some
vegetable mixtures the various components may have been frozen at different
times, etc. This means that the packaging date actually conveys only limited
information to the consumer.

Another confusing element is the fact to which reference has been made
above, that many products like peas, fried potatoes, shrimps, etc., are frozen
loose, e.g. in fluid-bed freezers, by direct immersion, in refrigerants, etc., and
subsequently bulk stored for custom packaging when required. In this case
also, the packaging date may be quite meaningless.

Manufacturers feel that labelling with packing date actually affects sales in
such a way that a product which has a somewhat earlier packing date than
another will stay in the retail cabinet and not be sold although the product
may be of perfectly good quality.

Nevertheless, consumers seem to feel quite strongly that a packaging date
should be given.

The author is of the opinion that the trade must try to comply with the
wishes - and whims - of the consumers since their satisfaction is the purpose of
their production efforts. However, since packaging data often has limited
meaning, it is conceivable that in the future a requirement might instead be

such time-temperature indicating systems as described in the section: Time-temperature integrators above.

Maximum permitted keeping time regulations

Authorities sometimes stipulate maximum keeping times permitted for various frozen products. As an example, Saudi Arabia required (1983) that frozen meat be sold not later than 9 months after freezing, and that imported frozen meat should be received in Saudi Arabia less than 3 months after freezing. Greece and the Federal Republic of Germany similarly have maximum permitted storage periods for some products.

It is easy to see that authorities feel a need for such regulations. The purpose of date marking is, of course, consumer protection, and it would be misleading if a manufacturer gave over-optimistically long keeping times for frozen products, although in principle, he might be free to do so.

However, it seems equally undesirable to have such keeping times stipulated by the authorities because this may remove any incentive from manufacturers to adopt improved procedures for obtaining longer keeping times, e.g. there might be little reason for manufacturers using the advantages which can be derived from the intelligent use of the many PPP-factors, some of which are mentioned above. Here it is well to keep in mind that in food processing, a very great many other factors are left to the discretion of the manufacturer, anyway. Inappropriate procedures, careless handling and unjustified claims must be covered by general requirements as regard good manufacturing practice, rules against misrepresentation, adulteration, sale of foods unfit for human consumption, etc., in food legislation and food control. Thus, a manufacturer giving an unusually long keeping time for a product could be required to document the data on which the claim is based and could be prosecuted as misleading the consumer in case of failure to do so satisfactorily.

It is interesting that the US Advisory Board on Military Personnel Supplies (AMBPS, 1982) proposed that it is desirable to develop a storage life index that makes the product's quality the criterion of acceptability instead of time it has been stored. It is to be hoped that both governments - and trade - will accept this view. At the same time, it must be granted that the implementation of this requires many and better data than are likely to be available at present.

Energy and Cost

Energy considerations

The question of energy consumption in the frozen food chain is often discussed. Questions regarding the freezing process and retail cabinets were reviewed above. As an example, Olabode *et al.* (1977), quoted by Hallström (1981), calculated that only by quite long storage times, especially at the retail

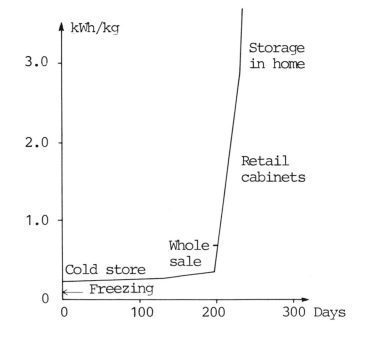

Fig. 151. Accumulated energy consumption in the freezer chain, based on Lorentzen (1977), and Porsdal Poulsen and Raahauge (1979).

Table 110. Power consumption per ton of the various links in the freezer chain. Estimated after Hallström (1981), Lorentzen (1977), Porsdal Poulsen and Raahauge (1979).

Freezing	90-150 kWh
Freezer storage warehouse (150 days)	310-250 kWh
Temperature maintenance during transport (transfer periods included)	30 kWh
Retail display cabinets (20 days)	2,500 kWh
Home freezer (10 days)	30,000 kWh

Table 111. Indexes for energy consumption at various types of temperature management in a freezer store. After Ashby *et al.* (1979).

	Index for energy use
−23°C constant	100
−23°C to -18oC fluctuating	92
−21.5°C to −18°C fluctuating	79
−18°C to −15°C fluctuating	76

level, would the distribution of foods in the frozen form be more energy consuming than their distribution in canned form, see also Bromander (1982). Table 110 and Fig. 151 give some consolidated data based on calculations by Hallström (1981), Lorentzen (1977), and Porsdal Poulsen and Raahauge (1979).

These as well as Table 88 reveal that, contrary to what is often assumed, comparatively litte energy is consumed in the freezing process and during wholesale storage and transport. Nevertheless, some savings might be achieved here if the predominant rigid attitude towards prevention of temperature fluctuations and maintenance of very cold temperatures should be somewhat relaxed; thus, Ashby *et al.* (1979) measured energy saving by maintaining various storage temperatures in a freezer store, as shown in Table 111.

The main energy consumption takes place in the retail cabinet. If home freezers are used, the energy consumption at this stage is also very high. However, where a home freezer is being operated at all, its power consumption takes place whether it contains more or less goods. Conversely, frozen foods are often purchased for immediate consumption and may then actually contribute slightly to energy saving in that they are often thawed in a home refrigerator.

Table 112. Direct energy use in the freezer chain for chopped meat in France, according to Grolee (1982).

	kJ/kg	kWh/t
Freezing	471	131
Freezer storage (1 month)	129	36
Retail cabinet (7 days)	158	44
Home freezer (3 days)	131	36

Table 113. Direct energy consumption in freezing and frozen distribution of pollack fillets, after CIRTA (1981).

	kJ/kg	kWh/t	%
Chilling raw material with ice	950	264	34
Washing, filletting and shimming	50	14	2
Freezing	256	71	9
Freezer storage (3 months)	389	108	14
Retail display, vertical glass door cabinets (7 days)	1008	280	37
Home freezer storage (3 days)	104	29	4
	2757	766	100

Grolee (1982) gives an account of direct and indirect energy requirements in the production, distribution and home preservation of frozen chopped meat. In so far as direct energy use is concerned, the data are given in Table 112.

In this calculation, very short storage times are assumed compared to those normally observed, cf. Tables 73, 74 and 76 and Figs 81 and 84. Meffert (1983) quotes considerably higher energy requirements in the various stages.

CIRTA (1981) arrives at the data reproduced in Table 113 for energy consumption in the various parts of the freezer chain for pollack in France.

One will notice that even when using freezer storage cabinets with side opening doors, and assuming only 7 days retail storage, retail display constitutes a very significant part of the total energy expenditure. Especially, if in this calculation one were to include storage times in retail cabinets of 20-30 days more closely resembling that actually found in surveys for the average product, cf. Table 77, and home storage periods of 20 days, cf. Fig.

Table 114. Cost of home storage of frozen foods in Dkr. per year. (1 Dkr. = 0.11 US$ = 0.14 ECU = 0.08 Eng. £ 0.33 DM). 200 kg is assumed frozen per year, average keeping time in home freezer 6 months.

Investment	Capital	Interest	Depreciation or loss	Annual cost
Freezer	2000	300	200	500
Product (av. 6 months)	1000	150	20	170
Operation				
Electricity, 1000 kWh/year, at 0.50				500
Freezer packaging, 200 kg at 0.30				60
				1230

111 and Table 94, more closely resembling that found in actual surveys, the energy expenditures in these links of the freezer chain become very large, indeed. Also, it is seen in Table 113 that even with the use of side opening vertical cabinets for retail display, the energy consumption in this part of the freezer chain becomes a factor of major importance.

Fine (1981) underlines the importance of energy conservation in the retail distribution of foods, stating that energy is the second highest operational expense, next to labour, in supermarket operations. This even applies to supermarkets using vertical glass door display cabinets for frozen foods.

Guldager (1982) describes a system with night lids, for both horizontal and vertical cabinets, heat recovery and controls by microcomputers assumed to bring about considerable energy savings in the management of freezer cabinets at the retail level.

Commercial versus home freezing

Recent years have seen an enormous increase in the number, see Fig. 1, and size of home freezers and the use of home freezing. Large companies have sprung up selling wholesale cartons with frozen foods, wholesale cuts of meat, etc., for use in home freezers. Consumers buy these at discount prices, take the foods home, freeze it if it is not already frozen, and keep it in their home freezers until they need it. It is likely that more research is required into the actual benefit for the consumers of this practice. In Table 114 a calculation is given of the cost which is actually incurred by the consumer.

When such costs as interest on the capital invested, depreciation of equipment, value of packaging material and a small but reasonable factor for unavoidable losses of food is added, it will be seen that the price advantage per kilogram of the product must be very substantial - or inflation very great

- before any net saving is achieved. It is likely that most consumers fail to realize the various cost components indicated here. Thus, very few households pay much attention to the electricity consumption of a home freezer, and few consider depreciation and interest on both equipment and the frozen product.

According to this type of calculation, a saving of Danish Kr. 6.15 (US$ 0.68) per kg must be realized to make the use of a home freezer economic for the consumer. If one assumes an inflation rate of 10 per cent, the necessary saving to make home freezing economic need only be Dkr. 5.65 (US$ 0.62) per kg.

Some may argue that this is an incorrect way of looking at the use of home freezers, since most consumers have and operate a home freezer to freeze left-overs, for convenience, etc. This means that cost of depreciation, interest and even power consumption will be incurred anyway. In that case, only interest on product cost, loss of product, and packaging has to be included, i.e. Danish Kr. 230.- per year. If the inflation rate again is set at 10 per cent, the necessary saving to make the use economic is only Danish Kr. 0.65 (US$ 0.07) per kg.

The latter is probably a more realistic calculation in many cases, but not when, as is the case in quite a few homes, consumers have resorted to having two home freezers to take care of extra purchases, etc.

The calculations do suggest that a significant saving in product cost is necessary to make it worthwhile to purchase products for home freezing. Secondly, if more purchases are anticipated than those that can be accommodated in the smallest size of home freezers, which is adequate for freezing left-overs, etc., costs start to increase. Further, if this practice is engaged in, it is more economic to use one large home freezer than to operate several smaller ones.

In the above, no allowance has been made for the extra work involved in preparing and packaging products for home freezing and the not insignificant task of keeping track of what is in the home freezer.

Conversely, the home freezer may make possible fewer purchasing trips and may thus save time and affect petrol consumption. Also, it will provide a better stock of foods on hand. At times, its operation may help space heating through heat from its condenser. In other cases, its operation will result in an extra load on air conditioners.

The above calculations suggest that this is an area which merits further study.

Thawing

Effect on quality

In many experiments regarding the quality of frozen foods, thawing has received comparatively little attention, although Fennema (1968) listed the thawing processes of considerable importance for the quality of frozen foods, second only to the frozen storage conditions. Obviously, no taste test can normally be carried out on a frozen product except on ice creams, sherbets, etc., without including the effect of thawing. The effect of thawing itself has been well studied in connection with the use of frozen foods in homes. Tables 115 - 117 give a few of many summaries of such findings.

In general, it seems that variations in thawing methods have a limited, but not insignificant, effect on total end product quality. However, in flesh products, e.g. beef or chicken, where texture is important, it seems that a slow thawing process at not too low a temperature is to be preferred. Thus, for instance, the practice of directly boiling or frying frozen fish fillets is convenient but not to be preferred from a quality point of view. The New Zealand lamb industry often labels their products: Defrost thoroughly before cooking.

Table 115. Effects of various methods for thawing beef frozen in 600 g pieces. Score 9 highest, 5 unsatisfactory. After Stoll, Dätwyler, Fausch and Neidhardt (1977).

	Methods and thawing times	Organoleptic evaluations			
		Colour	Form	Flavour	Texture
1.	Room temperature for 16 hrs.,then cooked for 2 hrs.	7.8	7.8	7.5	7.5
2.	Refrigerator 24 hrs., cooked 2 hrs.	7.3	7.8	8.0	7.3
3.	Cold tap water 3 hrs., cooked 2 hrs.	7.5	7.8	8.0	8.3
4.	Pressure cooked 45 mins., 8 mins. standing	8.0	7.8	7.5	6.8
5.	High pressure steam cooked 1 hr.	5.0	6.5	7.0	7.0
7.	Microwave 2100 W, 27 mins in glass dish with gravy	6.3	6.8	6.0	5.0

Table 116. Effects of various methods of thawing raspberries, frozen in 360 g packs, loose with sugar. Score: 9 highest, 5 unsatisfactory. After Stoll, Dätwyler, Fausch, and Neidhardt (1977).

Methods and thawing times			Organoleptic evaluations			
			Colour	Form	Flavour	Texture
1.	Refrigerator 4°C,	22 hrs.	6.9	5.8	7.6	6.5
2.	Room temperature 20°C	17 hrs.	6.5	5.1	6.9	5.5
3.	Tap water 12°C	1 hr.	6.6	5.8	5.5	5.7
4.	Convection oven,	1 hr.	6.6	5.8	5.5	5.7
5.	Pressure cooker,	3 mins.	7.3	6.9	7.1	5.6
6.	M-W oven 1300 W,	2 mins.	6.3	4.5	7.0	5.5
7.	M-W oven 1000 W,	2 mins.	7.3	6.1	7.6	7.0

Table 117. Effects of various freezing and thawing procedures on strawberries, frozen in 300 g packs, with loose sugar. R = Room temperature thawing, 17 hrs. at 20°C. M-W = Microwave thawing, 3 mins., 1000 W. Score: 9 highest, 5 unsatisfactory. After Stoll, Dätwyler, Fausch and Neidhardt (1977).

Freezing methods,	Tempe-rature	Thawing method and organoleptic evaluations							
		Colour		Form		Flavour		Texture	
		R	M-W	R	M-W	R	M-W	R	M-W
Still air, 20 hrs.	–20°C	4.7	7.3	4.5	7.3	5.3	7.5	4.0	6.5
Plate freezer, 2 hrs.	–38°C	5.0	7.3	4.8	7.3	6.3	7.7	5.0	6.5
Liquid nitrogen, 30 mins.	–70°C	5.3	7.3	5.3	7.3	6.5	7.0	5.3	6.3

It is generally assumed that the reason for the advantage of slower thawing for flesh foods is that of allowing time for diffusion to take place in the thawed tissue in order that water may return to its original position in the tissue.

On the other hand, products like strawberries, etc., seem to be best when thawed without delay and eaten as soon as practical thereafter, cf. Table 117 where microwave thawing consistently yields better results, a trend which is also seen in Table 116.

Vail, Jeffrey, Forney and Wiley (1943) tested thawing of beef and pork cuts. Generally, there was a slight preference for those thawed before cooking over those cooked direct.

Delaunay and Rosset (1981) describe various recommended thawing practices, especially as they are used in large scale food preparation.

Ke, Smith-Lall and Pond (1981) give the data reproduced in Table 118 as regards quality changes in thawing of Atlantic squid.

Table 118. Thawing characteristics for squid, acording to Ke, Smith-Hall and Pond (1981). (Details of grading system not given).

Methods	Thawing time (hour)	Thawed squid, % In grade A
Air 5-10°C	20-30	90
Air 15-20°C	8-15	70
Fresh water 10-15°C	6-10	61
Ice water 6-12°C	8-14	85

Kulp and Bechtel (1961) found rapid thawing superior to slow thawing for Danish pastry.

Industrial thawing

Industrial thawing appears to be an area which is in need of considerable research. Commercially, many frozen foods are thawed before sale. The purpose may be to sell them as fresh foods, or they may be further processed and refrozen, or they may be used as raw material for industrial processing.

These practices conflict with some old concepts, e.g. that thawed products should be consumed soon after thawing and must not be refrozen. It is hard to determine how this concept originated. It may actually be a result of a general human tendency of advocating against the unknown, e.g. freezing a food product twice. It may also be due to the observation that uncontrolled thawing often results in condensation and subsequent microbiological growth, resulting in product deterioration or outright decomposition before refreezing, reprocessing or use in the thawed state.

Carroll, Cavanaugh and Rohrer (1981) showed that repeated freezings and thawings have no adverse effect on meat structure except when thawing was carried out at room temperature for 24 hours. In the latter case, repeated freeze-thaw cycles resulted in extensive degradation.

Very sizeable industrial operations make use of frozen raw material for sale or reprocessing. Thus, fish is very frequently frozen in blocks at sea, thawed or partly thawed in port, filleted and refrozen or packaged and frozen as fish sticks, fish fingers, etc. Much frozen beef and lamb is shipped from overseas' suppliers to European users. It is practically always thawed, cut up, and sold in the fresh state. Few consumer complaints have ever been recorded on account of any of these processes. In the poultry industry it is not unusual to freeze a material twice before its subsequent use in a processed frozen product.

In the fishery industry, refreezing white fish may result in some loss in shelf life, cf. Fig. 58. For shrimp it is without any discernible effect, while frozen, cooked crab meat gets tough if refrozen.

Philippon (1981) describes experiments with freezing peaches directly as harvested. When they are to be used in further processing, they are tempered to -4°C and lye-peeled while still hard.

In the Danish ham industry it is customary to freeze hams for subsequent curing. Present practice is to use no more than 25% frozen raw material because of water binding considerations, the target is eventually to be able to use 100%.

This may not be attainable. Thus, Nusbaum (1979) found that the quality of ground beef is impaired if more than 50% consists of thawed meat. Conversely, up to this rate, he finds no adverse affect. Interestingly, he finds a considerably smaller cooking loss from thawed meat patties than from comparable unfrozen samples, provided the patties are thawed prior to cooking, cf. Table 59.

The production of salami is often based exclusively on the use of frozen raw material. Thawing takes place during the cutter operation.

As mentioned above, the Wiltshire bacon trade has, in periods, resorted to freezing the cured product for subsequent use in the thawed state, either as regular bacon or for further processing.

The Danish Meat Research Institute in Roskilde has found that processing Wiltshire bacon from frozen pork sides may result in a product of slightly improved quality but a 2% extra weight loss. Cooked ham prepared from frozen hams shows about 2% lower yield and a somewhat lighter colour of the finished product than if unfrozen raw material is used.

Research is indicated into methods of industrial thawing. One still observes the rather crude practice of simply spreading boxes of frozen products in the evenings around on tables or even on the factory floor in order that the produce may be thawed the next morning and ready for processing. Also, frequent use is made of thawing in running water, with a danger of leaching out some of the taste and nutrient factors. A lot of equipment has been developed specifically for industrial thawing. In general, much suggests that various types of air blast thawing equipment using controlled humidity and a carefully worked out temperature programme is the most effective for complete thawing, while dielectric or microwave thawing which can be very quick are preferable where only tempering is required. By tempering is meant increasing temperature of the frozen product to just below the freezing point, i.e. with an input of only about 1/4th of that energy which would be required for complete thawing. Many products can well be used in further manufacturing in this tempered condition.

One of the difficulties in thawing is that the thermal conductivity of non-frozen tissue is less than half of that of the frozen tissue. Therefore, thawing is somewhat slower than freezing. This makes it difficult to avoid overheating the surface of the food when it is bulk frozen. Thawing may result in temperature conditions as illustrated in Fig. 152.

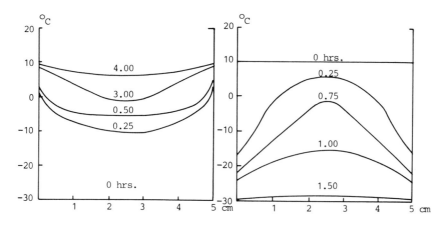

Fig. 152. Computed temperatures at various times during thawing, left, (+10°C) and freezing, right, (-32°C), of a 5.0 cm block of meat. Jason (1974).

Keeping quality of thawed products

Even the very early work by Plank, Ehrenbaum and Reuter (1916), demonstrated that thawed fish keeps just as well as unfrozen fish under the same conditions. Later, Reay (1933) tested fish which had been frozen according to different methods and then had been thawed and compared them with fresh fish. It was determined that the rate of bacterial growth on the surface of the fish was independent of whether the fish had been frozen or not and independent of freezing methods.

Simonsen (1961) froze and thawed a package of ground pork till it was completely thawed. He compared the thawed product with a non-frozen control sample and found the growth rate of surface bacteria as indicated in Fig. 153. No difference in keeping quality was observed. Many other investigations have arrived at similar conclusions. Only with uncontrolled thawing with heavy surface condensation was an adverse effect found. The same author (1963) reported similar results for thawed beef.

Bouillet, Dechene and Deroanne (1982) seem to confirm these findings with pork.

Many products are thawed in water, normally running water. When reasonable hygienic precautions are taken, no effect on microbiological quality is found.

Delaunay and Rosset (1981) give information about how thawing products for sale in the thawed condition has to be carried out in France. They point out, for instance, that regulations are very liberal with regard to thawing bread and bakery products, since no problems of hygiene seem to exist in this area.

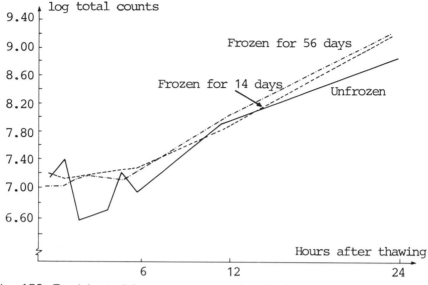

Fig. 153. Total bacterial count on ground pork, fresh and thawed. After Simonsen (1961).

The same authors mention that the keeping quality of thawed fruit and vegetables may be considerably affected by the fact that the breakdown of the cellulose cell walls may favour the activity of both endoenzymes and microorganisms; thus, the stability of products in these categories may be somewhat affected by an episode of freezing and thawing. On the other hand, freezing fruit and vegetables for processing later is very widespread. Further data on this aspect are desired.

Quality of thawed products

Boegh-Soerensen, The Danish Meat Products Laboratory (1977) carried out tests with trained taste panels. Rarely were they able to record any difference between an unfrozen meat or poultry product and one that had been frozen for a few days, cf. Table 18. For longer storage periods, frozen products will, of course, have undergone changes, i.e. as indicated by the TTT-curves given above.

Very little difference was found after different thawing methods for chicken parts by the Danish Meat Products Laboratory (1977), cf. Table 119.

Simonsen (1974) froze, thawed and refroze broilers up to four times. He recorded a - non-significant - quality improvement, cf. Table 120. Similarly, Baker *et al.* (1976) froze chicken broilers packaged in polyethylene bags at

Table 119. Effects of different thawing methods on organoleptic characteristics of broiler breasts (-5 dislike extremely, +5 like extremely). Danish Meat Products Laboratory, 1977.

Method		Taste	Texture	Juiciness
48 hours	4°C	-0.1	0.6	-1.0
24 hours	4°C	0.1	0.6	-1.4
20 hours	20°C	0.2	1.4	-0.9
40-45°C water		0.4	0.8	-1.6
Fried direct		-0.1	0.5	-1.3

Table 120. Organoleptic characteristics of broilers frozen once or several times. (-5 = dislike extremely, +5 = like extremely). After Simonsen (1974).

Frozen	Taste	Texture	Juiciness
Once	0.30	-0.36	-0.92
Twice	0.12	-0.14	-0.80
Three times	0.34	-0.26	-1.26
Four times	0.46	+1.16	-0.52

-18°C or -30°C. At 2-4 days interval they were thawed at room temperature for 7-8 hours to an internal temperature of 4°C. Random carcasses were removed for testing and the remainder refrozen up to five times. Taste panel scores showed that tenderness, flavour and overall acceptability of roasted breast meat were not affected by five refreezings. Other characteristics were not noticeably affected either.

Some data on the use of raw material in meat products are given in Tables 40, 41 and 59.

Ristić and Schön (1982) freezer-stored broiler chicken for extended periods, thawed them and prepared chicken parts therefrom which were then refrozen with satisfactory results. Total shelf life was unaffected by this process and, peculiarly enough, also independent of whether freezer storage at -12°C or -18°C was used.

In another experiment, Ristić (1982a) found that frozen chicken parts prepared from broilers freezer-stored at -12°C for 0-3 months kept for 3 months at -12°C. At -18°C, the corresponding shelf life of the frozen parts was 4 months. However, if the frozen broilers were freezer-stored 6-11 months, chicken parts prepared from the thawed carcasses showed a shelf life of only two months. Results were the same whether storage at -12 or -18°C was used.

Palumbo *et al.* (1982) demonstrated that the use for dry sausage of beef or pork which had been freezer stored for 90 weeks at -17.8°C resulted in a slight flavour loss but otherwise had no adverse effect at all.

Dyer *et al.* (1962) found no influence on quality as a result of thawing and refreezing but found that refreezing significantly reduced the shelf life of frozen cod. Later, experiments with cod, flounder and redfish by MacCallum *et al.* (1969) led to the conclusion that seafreezing and thawing, filleting or packaging and refreezing lead to products of a quality and shelf life not much inferior to that of the once frozen fish. Hansen (1969) obtained similar results for plaice, provided the fish were not iced more than five days prior to initial freezing.

Dyer *et al.* (1962), as quoted by Andersen, Jul and Riemann (1966), found some effect on the shelf life of refreezing fish as indicated in Fig. 58, and Nip and Moy (1981) found that prawns frozen for one month kept as well when thawed as the unfrozen product and were similar in flavour and acceptance.

Incorrect thawing may result in bacteriological problems; thus, the use of thawed beef has been implicated in Salmonella poisoning in the USA. Therefore, the Food Safety and Inspection Service of the USDA has recently issued specific regulations to cover meat thawing. Marriott *et al.* (1980a) suggest thawing ground beef at refrigerator temperatures for 24 hours to safeguard the product's bacteriological condition. Similarly, Marriot *et al.* (1980b) suggest that retail cuts of pork and lamb be thawed at refrigerator temperature, again to avoid undesirable bacterial growth.

One aspect of industrial thawing, which is not generally noted in home thawing, is that of drip loss, yield, etc. Above, it was mentioned that when large cuts of beef are thawed, one generally experiences a drip loss of a few per cent. If material for further manufacture is thawed, one will generally find a yield of a few per cent less than that obtained with an unfrozen product. Increased cooking losses of a few per cent would hardly be observed under household conditions but become factors of major economic importance in industrial operations.

Labelling thawed products

Since thawed products are looked upon with considerable suspicion, it is often suggested and sometimes even required by law that where such products are sold, this fact should be clearly indicated by appropriate labelling. From a technical and wholesomeness point of view, this may seem somewhat superfluous, since most foods have gone through many pre-sale processes which may affect product quality more than freezing and thawing. Yet, no one asks that all such processes be disclosed on the label.

In Finland, the freezing of previously frozen products is prohibited.

In the UK, the "Food Labelling Regulation, 1980", requires that any meat, including poultry meat or offal sold as such, which has been frozen and thawed shall be labelled "previously frozen - do not refreeze" unless the

Fig. 154. Designation on label and explanation, given in drop folder, regarding frozen fish, sold thawed, in an American supermarket, "Giant".

product is not prepackaged, in which case a notice of its history must be prominently displayed in the store.

In the Federal Republic of Germany, meat, which is sold fully or partly thawed, must be labelled, "Thawed - use immediately".

In the USA, the Department of Agriculture forbids that thawed poultry products be labelled "fresh" or "not frozen". One supermarket chain in the USA, "Giant Foods, Inc.", uses the designations shown in Fig. 154 to distinguish between products in this regard.

For meat, the USDA Meat and Poultry Inspection Regulation requires that thawed products be marked "Previously handled frozen for your protection. Refreeze or keep refrigerated".

According to Delaunay and Rosset (1981), thawed products may be sold in France. Local authorities in some areas require that it be clearly indicated that they have been frozen. Further, such products may never be designated fresh ("frais").

According to the EEC Labelling Directive which went into force on 1 January 1983, it may be required that frozen products, thawed and sold in the fresh state, be labelled accordingly. This interpretation has already been accepted in the UK (i.e. The Food Labelling Regulation, 1980).

From an enforcement point of view, requiring disclosure on the label of previous freezing causes considerable problems, since it is very difficult to determine if a product has been frozen. One method has been suggested for meat. Meat contains two types of glutamat-oxalacetat-transaminase, one found mainly inside the muscle cells and one mainly in the interstitial spaces. In freezing, some of the former is released. This may be verified analytically by electroforesis, but the procedure is somewhat cumbersome. Barbagli and Crescenzi (1981) suggest similarly a determination of cytochrome oxidase in extracts of meat. The content increases significantly after freezing and thawing. Gottesmann and Hamm (1982) have suggested that another mitochondriaenzyme, hydroxyacyl-CoA-Dehydrogenase be used (HADH). It may be determined by a simple dye-reaction. Obviously, none of the methods can be used for ground meat.

In considerations regarding permitted uses for previously frozen products it is overlooked that, in the meat control procedures of some countries, it is mandatory to freeze beef carcasses on which cysts are formed. Freezing destroys the parasite; the carcass is thawed and passed into the fresh meat trade without any indications of the freezing process it has undergone. No adverse reaction to this seems to have been reported.

Actual uses of thawed products

Very large quantities of meat are shipped from Argentina, Australia, and New Zealand to Europe as frozen carcasses, half carcasses, quarters, or boxed. Most of it is thawed, cut up and used in the fresh meat trade. No adverse effects of this practice has been reported.

As thawed meat is a regular part of the meat supply in some countries, thawed meat also very often forms the foundation for the production of frozen meat and meat dishes.

Meat is very often only partly thawed, i.e. tempered, and used for further processing in that form. The meat industry uses freezing to a considerable degree as a means of evening out supplies of raw material for curing, canning, etc.

Similarly, fish and shellfish are often frozen, thawed and sold in the fresh state. In New Zealand, according to Jarman (1982), rock lobsters are often frozen at sea, to be thawed and processed, often refrozen ashore.

Also, frozen fruit and vegetables are often frozen as raw material for subsequent thawing, or partial thawing, and uses in mixtures, prepared dishes, etc.

Philippon (1981) describes in detail how peaches may be bulk frozen for later processing and frequently for use in frozen fruit mixtures. The same subject is discussed by Philippon *et al.* (1981).

Freezing bread and bakery products is often used to avoid inconvenient working hours in bakeries and to meet peak demands. These products are normally sold as unfrozen products. Some problems of crust separation, etc., are sometimes seen.

Thawing is obviously an area where further studies are needed. Thawing in the home is fairly well studied; mainly the use of microwave ovens needs further elucidation. Industrial thawing, on the other hand, is still an area where many unsolved problems exist. Influence on yield, processing characteristics as well as quality and nutritive value need to be examined. Special attention must be given to the use of frozen products for sale in the unfrozen state as well as to their use in further processing, both freezing, curing and canning.

References

Aagaard, J. (1968). Fryselagring af fed fisk ("Freezer storage of fatty fish"). *Konserves og Dybfrost, 26* (7), 78-84.

ABMPS (1982). Predicting acceptance of frozen foods stored for extended periods. ABMPS Report No. 126. National Academy Press, Washington. 30 pp. (Mimeographed).

Amano, K. and Yamanaka, H. (1969). Meat colour regeneration in frozen tuna by irradiation. In: Kreuzer, R. (ed). Freezing and irradiation of fish. Fishing News (Books), London, 514-516.

Andersen, P.E.; Jul, M. and Riemann, H. (1966). Industriel Levnedsmiddelkonservering ("Industrial food preservation"). Teknisk Forlag, Copenhagen, Vol. II.

Andersen, I. and Andersen, P.E. (1979). Kostlære 1. Polyteknisk Forlag, Copenhagen.

Andover Laboratories (1983). Personal communication.

Ang, C.Y.W.; Chang, C.M; Frey, A.E. and Livingston, G.E. (1975). *J. Fd. Sci., 40,* 997-1003.

Anón, M.C. and Calvelo, A. (1980). *Meat Sci., 4,* 1-14.

Ashby, B.H.; Bennett, A.H; Bailey, W.A.; Moleeratanond, W. and Kramer, A. (1979). *Transactions of the ASEA, 22* (4), 938-943.

Aaström, S. and Löndahl, G. (1969). *Kyltekn. Tidsskr., 28* (6), 176-179.

Aurell, T.; Dagbjartsson, B.M. and Salomonsdottir, E. (1976). *J. Fd. Sci., 41,* 1165-67.

Baker, R.G.; Darfler, J.M.; Mulnix, E.J. and Nath, K.X. (1976). *J. Fd. Sci., 41,* 443-445.

Banks, A. (1935). *Rep. Fd. Invest. Bd.* (UK), 77-79.

Barbagli, C. and Serlupi Crescenzi, G. (1981). *J. Fd. Sci., 46,* 491-493.

Bech-Jacobsen, K. and Klinte, A. (1981). Plejehjemskøkkener - mad til hjemmeboende ("Institutional kitchens - food for home distribution"). Statens Levnedsmiddelinstitut, Copenhagen.

Bech-Jacobsen, K. and Klinte, A. (1981). Frossen færdigmad ("Frozen dishes"). Statens Levnedsmiddelinstitut, Copenhagen.

Bender, A.E. (1966). *J. Fd. Technol.* (UK), *1,* 261-289.

Bender, A.E. (1978). Food Processing and Nutrition. Academic Press, London.

Bengtsson, N; Liljemark, A.; Olsson, P. and Nilsson, B. (1972). An attempt to systemize time-temperature-tolerance (TST.T.) data as a basis for the development of time-temperature indicators. *Bull. Int. Inst. Refrig.* (Annexe 1972-2), 303-311.

272   *The Quality of Frozen Foods*

Bergh, F. (1949). *Kulde, 3* (3), 26-35.
Bevilacqua, A.E. and Zaritzky, N.E. (1980). *J. Fd. Technol.,* (UK), *15,* 589-597.
Birdseye, C. (1951). Theories and methods of quick freezing. In: "Refrigeration Data Book". ASRE, New York.
Blixt, K. (1983). Personal communication.
Boegh-Soerensen, L. (1967). Undersøgelse af forskellige kødfrysevarers tid-temperatur tolerance ("Investigations of the time-temperature-tolerance of various frozen meats"). Danish Meat Products Laboratory, Copenhagen. (Mimeographed).
Boegh-Soerensen, L. (1968). Holdbarheden af frosne pålägsvarer ("Keeping times for frozen pork cold cuts"). Danish Meat Products Laboratory, Copenhagen. (Mimeographed).
Boegh-Soerensen, L. (1975). Zur Haltbarkeit gefrorener Hähnchen und Hähnchenteilen, TTT- und PPP-Versuche ("TTT and PPP tests for broiler chicken and chicken parts"). *Fleischwirtschaft, 55* (11), 1587-1592.
Boegh-Soerensen, L. (1982a). TTT - The influence of packaging. In: Proceedings of the European Meeting of Meat Research Workers, *28,* Vol. I, 149.
Boegh-Soerensen, L. (1982b). Personal communication.
Boegh-Soerensen, L. and Bramsnaes, F. (1977). The effect of storage in retail cabinets on frozen foods. *Bull. Int. Inst. Refrig.* (Annexe 1977-1), 375-381.
Boegh-Soerensen, L. and Jensen, J. Hoejmark (1978). Påvirkningernes adderbarhed ved froslagring ("The additivity of the shelf life losses during freezer storage"). Report 42.02. Danish Meat Products Laboratory, Copenhagen. (Mimeographed).
Boegh-Soerensen, L. and Jensen, J. Hoejmark (1981). *Int. J. Refrig., 4* (3), 139-142.
Boegh-Soerensen L.; Jensen, J. Hoejmark and Jul, M. (1978). Konserveringsteknik ("Food preservation technology"). DSR, Copenhagen, Vol. 1.
Boggs, M.M.; Dietrich, W.C.; Nutting, M-D.F.; Olson, R.L.; Lindqvist, F.E., Bohart, G.S.; Neumann, H.J. and Morris, H.J. (1960). *Fd. Technol.* (US), *14* (4), 181.
Boldt, H. (1961). Den holdbarhedsmæssige betydning af temperaturvariationer i frost-disken ("The effect of temperature fluctuations in freezer display cabinets on product quality"). Danish Meat Products Laboratory, Copenhagen. 3 p. (Mimeographed).
Boström, R. (1982). *Scand. Refrig., 11* (3), 113-118.
Bouillet, D. M. and Deroanne, C. (1982). Frozen and refrigerated comminuted meat (beef and pork): microbial physio-chemical properties' evolution during storage. *Bull. Int. Inst. Refrig.* (Annexe 1982-1), 97-104.
Bowers, A.J.; Gardze, C. and Craig, J. (1979). *Poultry Sci., 58* (4), 830-834.
Bramsnaes, F. (1969). Quality and stability of frozen seafood. In: W.B. Van Arsdel, M.J. Copley and R.L. Olson (eds). Quality and Stability of Frozen Foods. Wiley Interscience, New York, 217-236.
Bramsnaes, F. and Soerensen, H.C. (1960). Vacuum-packaged frozen fatty fish. *Bull. Int. Inst. Refrig.,* 281-288.
Brenoe, C. and Jensen, J. Hoejmark (1981). Næringsstofanbefalinger og deres anvendelse ("Recommended intakes of nutrients and their application"). Publication No. 50, Statens Levnedsmiddelinstitut, Copenhagen.
Briley, G.C. (1980). *ASHRAE J., 22* (12), 30-32.
Brinch, J.; Boegh-Soerensen, L. and Green, E. (1982). Frostdiske: produktkvalitet, energiforbrug ("Retail freezer cabinets: product quality, energy consumption"). Manus. No. 236, Danish Meat Products Laboratory, Copenhagen. (Mimeographed).
Bromander, A.L. (1982). *Livsmedelsteknik* (S), *24* (7), 342-343.

Buchter, L. and Zeuthen, P. (1970). Nedfrysningshastighedens og frostlagringstidens indflydelse på oksekøds organoleptiske egenskaber ("The influence of freezing rate and freezer storage times on the organoleptic characteristics of beef"). Danish Meat Research Institute, Roskilde. (Mimeographed).

Bundesanstalt für Fleischforschung in Kulmbach (1981). Jahresbericht 1980. BAFF, Kulmbach.

Byrne, C.H. (1976). *Fd. Technol.* (US), *30* (6), 66-68.

Byrne, C.H. and Dykstra, K.G. (1969). Surveys of industry operating conditions and frozen product histories. In: W.B. Van Arsdel, M.J. Copley and R.L. Olson (eds). Quality and Stability of Frozen Foods. Wiley Interscience, New York, 331-344.

Callow, E.H. (1930). *Rep. Fd. Invest. Bd.* (UK), 71-74.

Calvelo, A. (1981). Recent studies on meat freezing. In: Developments in Meat Science-2. R. Lawrie (ed). Applied Science Publ., London, 125-158.

Carlson, C.J. (1969). Superchilling fish - a review. In: Kreuzer, R. (ed). Freezing and irradiation of fish. Fishing News (Books) Ltd., London.

Carroll, R.J.; Cavanaugh, J.R. and Rorer, F.P. (1981). *J. Fd. Sci., 46,* 1091.

Carseldine, A.J. and Weste, R.R. (1976). *CSIRO Fd. Res. Q., 36,* 41-45.

CEC (1979). Water content of frozen or deep frozen poultry - Examination of methods of determination: Guinea fowls and ducks. Information on Agriculture, No. 67, CEC, Brussels.

CEC (1980). Water content of deep-frozen poultry - Comparison of methods of determination. Information on Agriculture, No. 71. CEC, Brussels.

Chrystall, B.B.; Hagyard, C.J.; Gilbert, K.V. and Devine, C.E. (1982). Electrical stimulation and meat quality. *Bull. Int. Inst. Refrig.* (Annexe 1982-1), 47-51.

CIRTA (1981). Evaluation du coût énergétique de la conservation de filet de lieu noir par surgélation ("Estimation of power cost for preservation by quick freezing of Saithe fillets"). *Bull. Int. Inst. Refrig.* (Annexe 1981-4), 197-205.

Clark, D.B. (1983). Cholesteric liquid crystals as an infinitely reversible thermometer. (Paper prepared for the 16th International Congress of Refrigeration, Paris, 1983).

Clark, R.K. and Van Duyne, F. (1949). *Fd. Res., 14,* 221-230.

Clemmensen, J. and Zeuthen, P. (1974). Physical factors influencing the quality of wrapped pork sides during freezing. In: Meat freezing - Why and how?. C.L. Cutting (ed). MRI Symposium No 3, Meat Research Institute, Bristol.

Codex (1978). Report of the Working Group on Temperature and Quality of Quick Frozen Foods. Codex Alimentarius Commission, Joint FAO/WHO Food Standards Programme. FAO, Rome and WHO, Geneva.

Codex Alimentarius Commission (1983). Proposed draft international code for the handling of quick frozen foods during transport. ALINORM 83/37. Codex Alimentarius, Joint FAO/WHO Food Standards Programme, Rome.

Comité Interministériel de l'Agriculture et de l'Alimentation, Comité National du Codex Alimentarius (1980). Enquête sur le control des températures des denrées surgelées ("Inquiry into the control of temperatures of frozen foods"). Comité National du Codex Alimentarius, Paris.

Cook, G.A.; Love, E.F.; Vickery, J.R. and Young, W.J. (1926). *Austral J. Exp. Biol. Med. Sci., 3,* 15.

Connell, J.J. and Howgate, P.F. (1968). *J. Sci. Fd. Agric., 19,* 342-354.

Connell, J.J. and Howgate, P.F. (1969). *J. Sci. Fd. Agric., 20,* 469-476.

Connell, J.J. and Howgate, P.F. (1971). Consumer evaluation of fresh and frozen fish. In: Kreuzer, R. (ed). Fish inspection and quality control. Fishing News (Books) Ltd., London, 155-159.

Couden, H.N. (1969). Quality management. In: W.B. Van Arsdel, M.J. Copley and

274   The Quality of Frozen Foods

R.L. Olson (eds). Quality and Stability of Frozen Foods. Wiley Interscience, New York, 263-283.

Council of the Danish Frozen Food Industry (1982). Frozen Foods in Denmark 1981. Dybfrostraadet, Copenhagen. 13 p.

Craig, M. (1983). Personal communication.

Crigler, J.C. and Dawson, L.E. (1968). *J. Fd. Sci., 33,* 248-250.

CRIOC (1982). La Surgélation. Centre de Recherche et d'Information des Organisations de Consummateurs, Brussels.

Cutting, C.L. and Hollingsworth, D. (1974). Effect of freezing on nutritive value. In: Meat freezing - Why and how?. C.L. Cutting (ed). MRI Symposium No. 3, Meat Research Institute, Bristol.

Dalhoff, E. (1963). Holdbarhedsbestemmelser efter lagring ved frosttemperaturer af sliced bacon ("Determining shelf life of frozen sliced bacon"). Danish Meat Research Institute, Roskilde. (Mimeographed, confidential).

Dalhoff, E. and Jul, M. (1965). Factors affecting the keeping quality of frozen foods. In: Progress in Refrigeration Science and Technology. Pergamon Press, New York. Vol. I, 57-66.

Danish Meat Products Laboratory (1981). Report 1968-1980. Danish Meat Products Laboratory, Copenhagen.

Danish Meat Research Institute (1970). Personal communication.

Danmarks Statistik (1981). Statistisk Årbog 1981 ("Stastistical Yearbook 1981"). Danmarks Statistik, Copenhagen.

Dawson, L.E. (196). Stability of frozen poultry meat and eggs. In: W.B. Van Arsdel, M.J. Copley and R.L. Olson (eds). Quality and Stability of Frozen Foods. Wiley Interscience, New York, 143-167.

Delaunay, J. and Rosset, R. (1981). Produits surgélés ("Deep frozen products"). Informations techniques des services vétérinaires, No. 72-75. Paris.

Deutscher Kältetechnischer Verein. Arbeitsblätter ("Worksheets"). C.F. Müller, Karlsruhe.

Dietrich, W.C.; Nutting, M-D.F., Boggs, M.M. and Weinstein, N.E. (1962). *Fd. Technol.* (US), *16* (10), 123-128.

Dietrich, W.C.; Nutting, M-D.F.; Olson, R.L.; Lindqvist, F.E.; Boggs, M.M.; Bohart, G.S.; Neumann, H.J. and Morris, H. (1959). *Fd. Technol.* (US), *13* (2), 136-145.

Djupfrysningsbyraan (1970). Djupfrysningsbyrån om köpvanor i två varuhus ("The Frozen Food Institute on purchasing habits in two department stores"). Djupfrysningsbyrån, Stockholm.

Dybfrostinstituttet (1981). Dybfrost 1980 ("Deep frozen foods 1980"). Dybfrost instituttet, Copenhagen.

Dybfrostraadet (1982). Dybfrost ("Deep frozen foods") 1981. Dybfrostraadet, Copenhagen, 14 p.

Dybfrostraadet (1983). Personal communication.

Dyer, W.J.; Fraser, D.I.; Ellis, D.G.; Idler, D.R; MacCallum, W.A. and Laishley, E. (1962). Quality changes in stored refrozen cod fillets. *Bull. Int. Inst. Refrig.* (Annexe 1962-1).

Dyer, W.J. and Peters, J. (1969). Factors influencing quality changes during frozen storage and distribution of frozen products, including glazing, coating and packaging. In: Kreuzer, R. (ed). Freezing and irradiation of fish. Fishing News (Books) Ltd. London.

EEC (1976). Council Regulation No. 2967/76 of 23 November 1976 laying down common standards for the water content of frozen and deep-frozen chickens, hens and cocks.

ECE (1978). Joint ECE/Codex Alimentarius Group of Experts on Standardization of Quick Frozen Foods. Agri./WP.1/GE.3/R.52. ECE, Geneva. (Mimeographed).

Ege, R. (1978). Fødevare- og ernæringstabeller ("Food composition tables"). 3rd ed. Nyt Nordisk Forlag; Arnold Busck, Copenhagen.

Ellis, J.; Folkers, K.; Levy, M.; Takemura, K.; Shizukuishi, S.; Ulrich, R. and Harison, P. (1980). *Chem. Pathol. Pharm., 33* (2), 301-344.

Elsborg, L. (1982). Vitamin og mineralindtagelse ("Vitamin and mineral intake"). In: De ældres kost ("The diet of the elderly"). Husmandsforeningernes Husholdningsudvalg, Aarhus and Foreningen af jyske Landboforeninger, Skanderborg, p. 14.

Empey, W.A. (1933). *J. Soc. Chem. Ind., 52,* 230.

Empey, W.A. and Howard, A. (1954). *Fd. Preserv. Q., 14* (2), 33.

Engler, P.P. and Bowers, J.A. (1976). *J. Amr. Diet. Ass., 69,* 253-257.

Eskilson, P. (1978). *Scand. Refrig., 7* (2), 104-109.

Eskilson, P. (1979). *Scand. Refrig., 8* (2), 119-122.

Eskilson, P. (1983). Personal communication.

ETM (1983). Giganter i miniformat ("Miniature giants"). ETM, Stockholm.

Farquhar, J.W. (1981). Monitoring food handling by time/temperature devices. *Bull. Int. Inst. Refrig.* (Annexe 1981-4), 517-524.

Farquhar, J.W. (1982). *Int. J. Refrig., 5* (1), 50-54.

Fennema, O.R. (1968). *Proc. Meat Ind. Res. Conf.,* 1968, 109.

Fennema, O.R. (1975). Effects of freeze-preservation on nutrients. In: Nutritional evaluation of food processing. R.S. Harris and E.F. Karmas (eds). 2nd ed. AVI Publishing, Westport, Conn. (US), 244-288.

Fennema, O. (1982). Effect of processing on nutritive value of food: Freezing. In: Handbook of nutritive value of processed food. Vol. I: Food for human use. M. Rechcigl (ed.). CRC Press, Boca Raton, FL (US), 31-43.

Fennema, O.R.; Powrie, W.D. and Marth, E.H. (1973). Low temperature preservation of foods and living matter. Marcel Dekker, New York.

Fields, S.C. and Prusik, T. (1983). Time-temperature monitoring using solid-state chemical indicators. (Paper prepared for the 16th International Congress of Refrigeration, Paris, 1983).

Fine, H.E. (1981). Energy management in supermarkets. *Bull. Int. Inst. Refrig.* (Annexe 1981-4), 217-221.

Fleming, A.K. (1974). The New Zealand approach to meat freezing. In: Meat frezing -Why and how?. C.L. Cutting (ed) MRI Symposium No. 3, Meat Research Institute, Langford.

Folkers, D. and Spiess, W.E.L. (1982). *Temperatur Technik, 20* (2), 8-11.

Food and Nutrition Board (1980). Recommended daily allowances. National Academy of Sciences, Washington.

The Frozen Food Roundtable (1981). Code of recommended practices for the handling and merchandising of frozen foods. The Frozen Food Roundtable, Washington, D.C.

Fuster, C. and Prestamo, G. (1979). Conservation of frozen green beans. The effect of blanching on peroxidase, chlorophyll, pH, structure and organoleptic properties. In: Proceedings of the XVth International Congress of Refrigeration, Vol. III, 969-975.

Gortner, V.A.; Fenton, F.; Volz, E.E. and Gleim, E. (1948). *Ind. Eng. Chem., 40* (8), 1423-1426.

Gottesmann, P. and Hamm, R. (1982). *Fleischwirtschaft, 62* (10), 1301.

The Grocer (1982), (27. Feb.), 83.

276   The Quality of Frozen Foods

Grolee, J. (1982) Evaluation du coût énergétique du steak haché surgelé ("Estimation of power cost for quick-frozen minced steak"). *Bull. Int. Inst. Refrig.* (Annexe 1982-1), 379-388.
Guadagni, D.G. (1957a). *Fd. Technol.* (US), *11,* 171-176.
Guadagni, D.G. (1957b). *Fd. Technol.* (US), *11,* 471.
Guadagni, D.G. (1969). Quality and stability in frozen fruits and juices. In: W.B. Van Arsdel, M.J. Copley and R.L. Olson (eds). Quality and stability of frozen foods. Wiley Interscience, New York, 85-116.
Guadagni, D.G. and Nimmo, C.C. (1958). *Fd. Technol.* (US), *12* (6), 306.
Guadagni, D.G.; Nimmo, C.C. and Jansen, E.F. (1957a). *Fd. Technol.* (US), *11,* 33-42.
Guadagni, D.G.; Nimmo, C.C. and Jansen, E.F. (1957b). *Fd. Technol.* (US), *11,* 389-397.
Guadagni, D.G.; Nimmo, C.C. and Jansen, E.F. (1957c). *Fd. Technol.* (US), *11,* 633-637.
Guldager, F. (1982). *Scand. Refrig., 11* (2), 82.
Hallström, B. (1981). *Livsmedelsteknik* (S), *23* (4), 182-184.
Hammer, W.C.K. (1976). Frozen food - current international endeavours in regulating frozen food. *Bull. Int. Inst. Refrig.* (Annexe 1976-1), 227-234.
Hansen, I.; Hjarde, W.; Lieck, H.; Soendergaard, H. and Soerensen, A. (1976). *Husholdningsrådets tekniske medd.* (DK), *16*(4), 16-20.
Hansen, P. (1969). A comparison of the quality of frozen plaice fillets prepared from fresh or frozen raw materials. In: Kreuzer, R. (ed). Freezing and radiation of fish. Fishing News (Books) Ltd. London.
Hanson, H.L. and Fletcher, R.L. (1958). *Fd. Technol.* (US), *12* (1) 40-43.
Hanson, H.L.; Fletcher, L.R. and Campbell, A.A. (1957). *Fd. Technol.* (US), *13,* 339-343.
Harrison, D.L.; Hall, J.L.; Mackintosh, D.L. and Vail, G.E. (1956). *Fd. Technol.* (US), *10,* 104.
Hartzler, E.; Ross, W. and Willet, E.L. (1949). *Fd. Res., 14,* 15-24.
Hawkins, A.E.; Pearson, C.A. and Raynor, D. (1973). *Proceedings of the Institute of Refrigeration* (UK), *69.*
Heiss, R. (1936). *Z. angew. Chem., 49,* 17.
Heiss, R.(1942) *Z. gesamte Kälteind., 49,* 131-136, 142-145.
Heldman, D.R. (1982). *Fd. Technol.* (US), *36* (2), 92-95.
Helms, P. (1978). Næringsstoftabeller. Lægeforeningens Forlag, Copenhagen.
Helms, P. (1981). *Ernæringsnyt* (DK), (5), 2.
Hicks, E.W. (1944). *J. Council Scient. Ind. Res., 17* (2), 111-114.
Hiner, R.L.; Madsen, L.L. and Hankins, O.G. (1945). *Fd. Res., 10,* 312.
Hjarde, W.; Laier, G.; Soendergaard, H.; Soerensen, A.; Christensen, B.; Mortensen, A.B. and Kousgaard, K. (1981). *Husholdningsrådets tekniske medd.* (DK), *21* (4), 19-32.
Honikel, K.O. and Fischer, C. C. (1982). *Fleischwirtschaft,* 60 (9), 1709-1714.
Hustrulid, A; Winter, J.D. and Noble, I. (1949). *Refrig. Eng., 57,* 38-41, 88.
Immerman, A.M. (1981). Vitamin B_{12} status on a vegetarian diet. In: *Wld. Rev. Nutr. Diet, 37,* 38-54
Jarman, N.E. (1982). The importance of appropriate refrigeration to the New Zealand fishing industry. *Bull. Int. Inst. Refrig.* (Annexe 1982-1), 141-145.
Jason, A.C. (1974). *IFST Proc., 7* (3), 146-157.
Jensen, J. Hoejmark (1981). Meat in international dietary patterns. In: Meat in nutrition and health. K.R. Franklin and P.N. Davis(eds). National Live Stock and Meat Board, Chicago, 35-63.
Jul, M. (1944/45). *Ingeniøren* (DK), (79, 85) 1944, (3, 11, 16) 1945.

Jul, M. (1960). Observations on the calculation of keeping quality of frozen foods. *Bull. Int. Inst. Refrig.* (Annexe 1960-3), 377-405.

Jul, M. (1969). Quality and stability of frozen meats. In: W.B. Van Arsdel, M.J. Copley and R.L. Olson (eds). Quality and stability of frozen foods. Wiley Interscience, New York, 191-216.

Jul, M. and Dalhoff, E. (1961). *Kulde* (DK), *15* (1, 2).

Jul, M. and Zeuthen, P. (1981). Quality of pig meat for fresh consumption. *Prog. Fd. Nutr. Sci.,* 4 (6), 1-132.

Kahn, L.N. and Livingston, G.E. (1970). *J. Fd. Sci., 35,* 349-351.

Ke, P.J.; Smith-Lall, B. and Pond, G.R. (1981). Quality enhancement of Canadian Atlantic squid (Illex illecebrosus) by various low temperature operations. *Bull. Int. Inst. Refrig.* (Annexe 1981-4), 513-516.

Kemp, J.D.; Montgomery, R.E. and Fox, J.D. (1976). *J. Fd. Sci., 41* (1), 1-3.

Klose, A.A.; Pool, M.F., Campbell, A.A. and Hanson, H.L. (1959). *Fd. Technol.* (US), *13* (9), 477.

Klose, A.A.; Pool, M.F. and Lineweaver, H. (1955). *Fd. Technol.* (US), *9,* 372.

Knutsson, P.E. (1980). Personal communication.

Kondrup, M. and Boldt, H. (1960). The influence of the freezing rate upon the quality of frozen meat and poultry. *Bull. Int. Inst. Refrig.* (Annexe 1960-3), 309-330.

Kotschevar, L.H. (1955). *J. Amr. Diet. Ass., 31,* 589-596.

Kozlowski, A. von (1977). Is it necessary to blanch all vegetables before freezing? *Bull. Int. Inst. Refrig.* (Annexe 1977-1), 227-238.

Kramer, A. (1979). *Fd. Technol.* (US), *33* (2), 58-61, 65.

Kramer, A.; Bender, F.E. and Sirivichaya, S.(1980). *Int. J. Refrig., 3* (6), 353-359.

Kramer, A. and Farquhar, J.W. (1977). Experience with TT-ii ....s. *Bull. Int. Inst. Refrig.* (Annexe 1977-1), 401-406.

Kramer, A.; King, R.L. and Westhoff, D.C. (1976). *Fd. Technol.* (US), *30* (1), 56-57, 60-62.

Kulp, K. and Bechtel, W.G. (1961). *Fd. Technol.* (US), *15* (5), 273-275.

Labuza, T.P. (1982). Shelf-life dating of foods. Food and Nutrition Press, Inc., Westport, 500 p.

Lea, C.H. (1930). Changes in fats during storage. *Rep. Fd. Invest. Bd.* (UK) 30-33.

Lea, C.H. (1935). *Rep. Fd. Invest. Bd.* (UK), 79-82.

Lee, F.A; Brooks, R.F.; Pearson, A.M.; Miller, J.I. and Volz, F. (1950). *Fd. Res., 15* (1), 8-15.

Lee, F.A.; Brooks, R.F.; Pearson, A.M.; Miller, J.I. and Wanderstock, J.J. (1954). *J. Amr. Diet. Ass., 30,* 351-354.

Lehrer, W.P.; Wiese, A.C.; Harvey, W.R. and Moore, P.R. (1951). *Fd. Res., 16,* 485.

Lenartowicz, W; Plocharski; Zbroszczyk, J. and Piotrowski, J. (1979). The effects of freezing method on the quality of frozen fruits. In: Proceedings of the XVth International Congress of Refrigeration, Vol. III, 901-908.

Lindeloev, F. (1977). Chemical reactions in frozen, cured meat products. *Bull. Int. Inst. Refrig.* (Annexe 1977-1), 153-165.

Lindeloev, F. (1978) Frostlagring af saltede produkter ("Freezer storage of salted products"). Laboratoriet for Levnedsmiddelindustri, Danmarks Tekniske Højskole, Copenhagen.

Livsmedelsteknik (S) (1980), *22* (8), 361.

Lorentzen, G. (1962). *Kjøleteknikk og Fryserinæring,* (5), 125.

Lorentzen, G. (1971). *Scand. Refrig.,* 1 (1), 5-11

Lorentzen, G. (1977). Personal communication.

Lorentzen, G. (1978). *Scand. Refrig., 7* (2), 67-70.

Lorentzen, G. and Roesvik, S. (1959). Noen undersøkelser vedrørende frysning av kjøtt ("Some experiments regarding freezing of meat"). Norges Tekniske Høgskole - Trykk, Trondheim.

Lorentzen, J. (1981). Personal communication.

Lowry, P.D. and Gill, C.O. (1982). Microbial considerations in cold storage of meat. *Bull. Int. Inst. Refrig.* (Annexe 1982-1), 91-96.

Lüthje, H. (1981). Tab af B-vitaminer i kødfrysevarer ("Losses of B-vitamins in frozen meats"). Manus. No. 229. Danish Meat Products Laboratory, Copenhagen. (Mimeographed).

Löndahl, G. (1977). How to maintain a sufficiently low temperature in frozen food distribution. *Bull. Int. Inst. Refrig.* (Annexe 1977-1), 355-362.

Löndahl, G. and Danielsson, C.E. (1972). Time-temperature-tolerances for some fish and meat products. *Bull. Int. Inst. Refrig.* (Annexe 1972-2), 295-301.

Löndahl, G. and Nilsson, T.E. (1978). *Int. J. Refrig., 1* (1), 53-56.

Löndahl, G. and Aaström, S. (1972). Glazing of frozen foods. *Bull. Int. Inst. Refrig.* (Annexe 1972-2), 287-293.

McBride, R.L. and Richardson, K.C. (1979). *J. Fd. Technol.* (UK), *14,* 57-67.

McCallum, W.A. and Chalker, D.A. (1967). *J. Fish Res. Board Canada, 24,* 127-144.

McCallum, W.A.; Jaffray, J.I.; Chalker, D.A. and Idler, D.R. (1969). In: Kreuzer, R. (ed). Freezing and irradiation of fish. Fishing News (Books) Ltd., London.

McCallum, W.A.; Laishley, E.J.; Dyer, W.J. and Idler, D.R. (1966). *J. Fish. Res. Board Canada, 23,* 1063.

McColloch, R.J.; Rice, R.G.; Bandurski, M.B. and Gentili, B. (1957). *Fd. Technol.* (US), *11* (8), 444.

McIntire, J.M. (1943). *J. Nutr., 25,* 143.

Manske, W.J. (1983). The application of controlled fluid migration to temperature limit and time temperature integrators. (Paper prepared for the 16th International Congress of Refrigeration, Paris, 1983).

Marriott, N.G.; Garcia, R.A.; Pullen, J.H. and Lee, D.R. (1980). *J. Fd. Protec., 43* (3), 180-184.

Marriott, N.G.; Garcia, R.A.; Kurland, M.E. and Lee, D.R. (1980). *J. Fd. Protec., 43* (3), 185-189.

Marsh, B.B. (1966). Meat Tenderness. In: Annual Research Report 1965-66. Meat Industry Research Institute of New Zealand, Hamilton, 1966, 14-18.

Mascheroni, R.H.; Ann, M.C. and Calvelo, A. (1981). *Meat Sci., 5,* 457-472.

Mecchi, E.P.; Pool, M.F.; Behman, G.A.; Hamachi, M. and Klose, A.A. (1956). *Poultry Sci., 35,* 1238.

Meyer, B.; Mysinger, M. and Burckley, R. (1963). *J. Agr. Fd. Chem., 11* (6), 525-527.

Meffert, M.F.Th. (1983). Tiefkühlketten in Theorie und Praxis ("Deep frozen food chain in theory and pratice"). Sprenger Institute, Wageningen. Mimeographed.

Michener, H.D; Thompson, P.A. and Dietrich, W.C. (1960). *Fd. Technol.* (US), *14,* 290-294.

Middlehurst, J; Richardson, K.C. and Edwards, R.A. (1972). *Fd. Technol. in Australia, 24* (11), 560-571.

Miller, W.O. and May, K.N. (1965). *Fd. Technol.* (US), *19* (7), 147.

Mills, A.E. (1981). Time-temperature management throughout storage and transportation with a thermograph engineered to record temperature variations for periods from 4 to 90 days. *Bull. Int. Inst. Refrig.* (Annexe 1981-4), 487-490.

MIRINZ (1981). Annual Report. Meat Industry Research Institute, New Zealand, Hamilton.

Mistry, V.K. and Kosikowski (1983). *J. Fd. Protec., 46* (1), 52-57.

Moleeratanond, W; Ashby, B.H.; Kramer, A; Berry, B.W. and Lee, W. (1981). *J. Fd. Sci., 46* (3), 829-833 and 837.

Moleeratanond, W; Kramer, A; Ashby, B.H.; Bailey, W.A. and Bennett, A.H. (1979). *Int. J. Refrig., 2* (6), 221.

Moran, T. (1932). *Rep. Fd. Invest. Board,* (UK), 22-26.

Morris, T.N. and Barker, J. (1932). *Rep. Fd. Invest. Board,* (UK), 92-94.

Morrison, M. (1974). *J. Fd. Technol.* (UK), *9,* 491-500.

Mullenax, D.C. and Lopez, A. (1975). *J. Fd. Sci., 40,* 310-312.

Munter, A.M.; Byrne, C.H. and Dykstra, K.G. (1953). *Fd. Technol.* (US), *7,* 356.

Nestorov, N. and Kozhuharova (1975). Studies on the changes in thiamin and retinol content in pig liver and muscles after sharp freezing and long cold storage. In: Proceedings of the European Meeting of Meat Research Workers, *20,* 269-71.

Nicol, D.L. and Spencer, R. (1960). Some factors affecting the loss of quality of frozen fish in distribution. *Bull. Int. Inst. Refrig.* (Annexe 1960-3).

Nielsen, L.M. and Carlin, A.F. (1974). *J. Amr. Diet. Ass., 65* (1), 35-40.

Nilsson, R. and Hällsaas, H. (1973). *Livsmedelsteknik* (S), *15* (1), 26.

Nip, W.K. and Moy, J.H. (1981). *J. Fd. Process. Preserv., 5,* 207-213.

Noble, I. (1970). *J. Amr. Diet. Ass., 56* (3), 225-228.

Nusbaum, R.P. (1979). *Recip. Meat Conf. Proc., 32,* 23-40.

Nusbaum, R.P.; Sebranek, J.G; Topel, D.G. and Rust, R.E. (1983). *Meat Sci., 8,* 135-146.

Olley, J. (1976). Temperature indicators, temperature integraters, temperature function integraters and the food spoilage chain. *Bull. Int. Inst. Refrig.* (Annexe 1976-1), 125-132.

Olley, J. (1978). *Int. J. Refrig., 1* (2), 81-83.

Olson, R.L. and Dietrich, W.C. (1969). Quality and stability of frozen vegetables. In: W.B. Van Arsdel, M.J. Copley and R.L. Olson (eds). Quality and stability of frozen foods. Wiley Interscience, New York, 117-141.

Olsson, P. and Bengtsson, N. (1972). Time-temperature conditions in the freezer chain. Report No. 309. SIK - The Swedish Food Institute, Gothenburg.

Ordynsky, G. (1964). *Tiefkühl-Praxis, 5,* 7.

Palmer, A.Z; Brady, D.E.; Naumann, H.D. and Tucker, L.N. (1953). *Fd. Technol.* (US), *7,* 90.

Palumbo, S.A.; Miller, A.J.; Gates, R.A. and Smith, J.L. (1981). *Fleischwirtschaft, 61* (6), 933.

Pearson, A.M.; Burnside, J.E.; Edwards, H.M.; Glasscock, R.S.; Gunha, T.J. and Novak, A.P. (1951). *Fd. Res., 16* (1), 85-87

Pearson, A.M.; West, G.G. and Luecke, R.W. (1959). *Fd. Res., 24,* 515-519.

Pedersen, Randi (1982). Immersion chilled fresh chickens. Organoleptic characteristics, shelf life and determination of water uptake. Manus. No. 230. Danish Meat Products Laboratory, Copenhagen. (Mimeographed).

Pence, J.W. (1969). Quality and stability of bakery products. In: W.B. Van Arsdel, M.J. Copley and R.L. Olson (eds). Quality and stability of frozen foods. Wiley Interscience, New York, 169-190.

Persson, P.0. and Nilson, T. (1967). *Livsmedelsteknik* (S), *9,* 76-83.

Philippon, J. (1981). *Rev. Gen. du Froid,* (3), 127-136.

Philippon, J. and Rouet-Mayer, M.A. (1978). Congélation sans traitement préalable de pêches destinées a l'industrie. *Bull. Int. Inst. Refrig.* (Annexe 1978-2), 101-110.

Philippon, J.; Rouet-Mayer, M.A.; Gallet, D. and Herson, A. (1981). Congélation, entreposage, congélation et reutilisation des pêches entières en conserverie de fruits au sirop ("Freezing, freezer storage and reuse of whole peaches in preserving fruits in sirup"). CNRS, Meudon.

Philippon, J.; Ulrich, R. and Zeuthen, P. (1982). Compte rendu de la Journée d'Etude international sur le Blanchiment des Fruits et Légumes surgelés. Etat des Connaissances actuelles - Perspectives d'Avenir. CNRS, Paris.

Plank, R.; Ehrenbaum, E. and Reuter, K. (1916). Die Konservierung von Fischen durch das Gefrierverfahren ("Fish freezing"). Zentral-Einkaufgesellschaft. Berlin.

Plessey (1972). Products data, P.D. 2076. February 1972. Plessey Australia Pty. Ltd., Villawood, NSW, Australia.

Porsdal Poulsen, K. and Raahauge, L. (1979). *Scand. Refrig., 8* (4), 325.

Rasmussen, C.L. and Olson, R.L. (1972). *Fd. Technol.* (US), *26* (12), 32-47.

Reay, G.A. (1933). *Rep. Fd. Invest. Board,* 167-178.

Redstrom, R.A. (1971). Practices in the use of home freezers. USDA Home Economics Report No. 38 (June), USGPO, Washington.

Reid, D.S. (1983). *Fd. Technol.* (US), *37,* (4), 110.

Reindl, B.; Grossklaus, R. and Busse, H. (1983). *Fleischwirtschaft, 63* (7), 1193-1196.

Richardson, L.R.; Wilkes, S. and Ritchey, S.J. (1961). *J. Nutr., 73,* 363-368.

Riordan, P.B. (1976). *I. J. Agric. Res., 15,* 185-195

Ristić, M. (1975). *Die Kälte, 28* (4), 133-138.

Ristić, M. (1980a). *Fleischwirtschaft, 60* (9), 1607.

Ristić, M. (1980b). *Fleischwirtschaft, 60* (10), 1894-1895.

Ristić, M. (1982a). *Die Fleischerei, 33,* (3), 178-180.

Ristić, M. (1982b). *Mitteilungsblatt der BAFF Kulmbach,* No. 76, 5042-5047.

Ristić, M. and Schön, L. (1982). Auftauen und erneuter Einfrieren nach erfolgter Zerlegung von kurz- und langfristig eingelagertem Schlachtgeflügel ("Thawing and refreezing after cutting-up in parts of poultry frozen for shorter or longer periods"). In: Kulmbacher Woche 1982 ("Kulmbach Week 1982"). Bundesanstalt für Fleischforschung, Kulmbach.

Ristić, M.; Schön, L. and Tawfik, E.S. (1976). *Arch. Geflügelkunde, 40* (4), 130-135.

Ristić, M. and Tawfik, E.S. (1975). *Arch. Geflügelkunde, 39* (5), 176-179.

Rogowski, B. (1974). *Fleischwirtschaft, 54* (10), 1010-1012.

Rogowski, B. (1976). *Fleischwirtschaft, 56* (2), 250-252.

Ronsivalli, L.J. and Gorga, C. (1981). Quality assurance in chilled and frozen seafoods. *Bull. Int. Inst. Refrig.* (Annexe 1981-4), 497-503.

Rose, R.E. (1981). Quality assurance in the cold chain - use of an enzymatic time/temperature monitor during transport and storage of refrigerated fish. *Bull. Int. Inst. Refrig.* (Annexe 1981-4), 483-486.

Sanderson-Walker, M. (1979a). *Int. J. Refrig., 2* (2), 93-96.

Sanderson-Walker, M. (1979b). *Int. J. Refrig., 2* (2), 97-101.

Sandström, B. (1974). Energiförbrukning vid konservering och lagring av livsmedel ("Energy consumption by preservation and storage of foods"). Report No. 352. SIK - The Swedish Food Institute, Gothenburg. (Mimeographed).

Sebranek, J.G. (1982). *Fd. Technol.* (US), *36* (4), 120-127.

Schubert, H. (1977). Criteria for application of T-T indicators to quality control of deep frozen food products. *Bull. Int. Inst. Refrig.* (Annexe 1977-1), 407-423.

Schwimmmer, S. and Ingraham, L.L. (1959). *Ind. Eng. Chem., 47* (6), 1149.

Sharp, A.K. and Irving, A.R. (1976). *Fd. Technol. in Australia, 28,* 295.

Shestakov, Y.U.M. (1976). *Veterinariya,* (3), 107-108.

Simonsen, B. (1961). Frysningens indflydelse på bakterievæksten i optøet hakket svinekød ("The influence of freezing on microbiological activity in thawed, ground pig meat"). Danish Meat Products Laboratory, Copenhagen. (Mimeographed).

Simonsen, B. (1963). Holdbarheden af optøet hakket kød i sammenligning med ikke-frosset ("Shelf life of thawed ground beef compared with the non-frozen product"). Danish Meat Products Laboratory, Copenhagen. (Mimeographed).

Simonsen, B. (1974). Kvaliteten af optøede og genfrosne kyllinger ("The quality of thawed and refrozen broiler chickens"). Danish Meat Products Laboratory, Copenhagen. (Mimeographed).

Singh, R.P. (1976). Computer simulation of food quality during frozen food storage. *Bull. Int. Inst. Refrig.* (Annexe 1976-1), 197-203.

Spiess, W.E.L. (1977). Personal communication.

Spiess, W.E.L.; Wolf, W; Wien, K.J. and Jung, G. (1977). Temperature maintenance during the local distribution of deep frozen foods. *Bull. Int. Inst. Refrig.* (Annexe 1977-1), 367-374.

Spiess, W.E.L. (1980). The shelf life of deep frozen food products. *Confidential draft,* made available for the use of the Codex Alimentarius Working Group on Frozen Foods, (Reprinted in excerpts with permission from the author).

Spinelli, J.; Pelroy, G. and Miyauchi, D. (1969). Quality indices that can be used to assess irradiated seafoods. In: Kreuzer, R. (ed). Freezing and irradiation of fish. Fishing News (Books) Ltd., London, 425-430.

Statens Levnedsmiddelinstitut (1981). Næringsstofanbefalinger og deres anvendelse ("Nutrient intake recommendations and their use"). Statens Levnedsmiddelinstitut, Copenhagen.

Steinbuch, E; Spiess, W.E.L. and Grünewald, Th. (1977). The long term storage of deep frozen green beans. *Bull. Int. Inst. Refrig.* (Annexe 1977-1), 239-247.

Stoll, K; Dätwyler, D.; Fausch, M. and Neidhardt, T. (1977). Thawing of frozen foods by different methods. *Bull. Int. Inst. Refrig.* (Annexe 1977-1), 393-399.

Storey, R.M. and Owen, D. (1982). Personal communication.

Strachan, P.W. (1983). *IFST Proc.,* 16 (2), 65-74.

Sundhedsprioriteringsudvalget (DK) (1977). Rapport-samling vedrørende forebyggelse. ("Collection of reports regarding prevention of disease"). Bilag til betænkning 809/1977. Indenrigsministeriet, Copenhagen.

Symons, H.W. and Cutting, C.L. (1977). Developments in food quick freezing, 1971-75. In: Proceedings of the XIVth International Congress of Refrigeration, Vol. III, 864-868.

Tanaka, K. and Tanaka, T. (1956). *J. Tokyo Coll. Fish., 42,* 80-82.

Tatterson, I. and Windsor, M.L. (1972). *J. Sci. Fd. Agric., 23,* 1253-1259.

Tressler, D.K. and Evers, C.F. (1957). The freezing preservation of foods. AVI Publishing Co., Westport, Vol. I-II.

Tuchschneid, M.W. (1936). Die kältetechnologische Verarbeitung schnellverderblicker Lebensmittel ("Processing perishable foods by refrigeration"). Kirchhain.

Tuchschneid, M.W. and Emblik, E. (1959). Die Kältebehandlung schnellverderblicher Lebensmittel ("Chilling and freezing perishable foods"). Brücke-Verlag Kurt Schmersow, Hannover.

UKAFFP (1978). Code of recommended practice for the handling of quick frozen foods. The UK Association of Frozen Food Producers, London.

USDA (1980). Food and nutrient intakes of individuals in 1 day in the United States, Spring, 1977. Nationwide Food Consumption Survey 1977-78. Preliminary Report No. 2, USDA, Washington.

Ulrich, R. (1981). *Rev. Gen. du Froid, 71,* (7/8), 371-389.

Vail, G; Jeffery, M; Forney, H. and Wiley, C. (1943). *Fd. Res., 8,* 337-342.

Van Arsdel, W.B. (1957). *Fd. Technol.* (US), *11* (1), 28.

Van Arsdel, W.B. and Guadagni, D.G. (1959). *Fd. Technol.* (US), *13* (1), 14.

Van Arsdel, W.B. (1969). Estimating quality change from a known temperature history. In: W.B. Van Arsdel, M.J. Copley and R.L. Olson (eds). Quality and stability of frozen foods. Wiley Interscience, New York, 237-262

Van Arsdel, W.B.; Copley, M.J. and Olson, R.L. (1969). Quality and stability of frozen foods. Wiley Interscience, New York.

Van den Berg, L. (1966). *Food in Canada, 26* (5), 37-38.

Ware, M.S. (1973). *Quick Froz. Foods Int., 14* (4), 158.

West, L.C.; Titus, M.C. and Van Duyne, F.O. (1959). *Fd. Technol.* (US), *13* (6), 323-327.

Westerman, B.D.; Vail, G.E., Tinklin, G.L. and Smith, J. (1949). *Fd. Technol.* (US), *3,* 184.

Westerman, B.D. *et al.* (1952). *J. Amr. Diet. Ass., 28,* 49-52.

Westerman, B.D.; Oliver, B. and Mackintosh, D.L. (1955). *J. Agric. Fd. Chem., 3* (7), 603-605.

Winger, R.J. (1982). The effect of processing variables on the storage stability of frozen lamb. *Bull. Int. Inst. Refrig.* (Annexe 1982-1), 75-82.

Winger, R.J. and Pope, C.G. (1981). *Int. J. Refrig., 4* (6), 335-339.

Wismer-Pedersen, J. and Sivesgaard, A. (1957). *Kulde, 5,* 54.

Woltersdorf, W. (1982a). Personal communication.

Woltersdorf, W. (1982b). Energiesparung beim Gefrieren und durch Warmzerlegung ("hot boning") ("Energy saving by freezing and hot boning"). In: "Kulmbacher Woche". BAFF, Kulmbach.

Woolridge, W.R. and Bartlett, L.H. (1942a). *Modern Refrig.*

Woolridge, W.R. and Bartlett L. H. (1942b). *Mechan. Refrig.,* (Sept.)

Zacharias, R. and Bognar, A. (1975). *Ernährungs-Umsch., 22* (2), 36-41.

Zaehringer, M.V.; Bring, S.V.; Richard, C.A. and Lehrer Jr., W.P. (1959) *Fd. Technol.* (US), *13* (6), 313-317.

Zaritzky, N.E.; Ann, M.C. and Calvelo, A. (1982). *Meat Sci., 7,* 299-312.

Zipser, M.W.; Kwon, T. and Watts, B.M. (1964). *J. Agric. Fd. Chem., 12,* 105.

Zipser, M.W. and Watts, B.M. (1967). *Agric. Fd. Chem.,* 15, 80.

FIGURE REFERENCES to PAGE NUMBERS

Fig. No.	Page No.	Fig. No.	Page No.	Fig. No.	Page No.	Fig. No.	Page No.	Fig. No.	Page No.	Fig. No.	Page No.
1 -	3	31 -	60	61 -	131	91 -	169	121 -	217	151 -	256
2 -	6	32 -	61	62 -	132	92 -	170	122 -	218	152 -	265
3 -	9	33 -	62	63 -	133	93 -	171	123 -	219	153 -	266
4 -	9	34 -	63	64 -	134	94 -	173	124 -	221	154 -	269
5 -	10	35 -	66	65 -	135	95 -	176	125 -	223		
6 -	10	36 -	67	66 -	138	96 -	177	126 -	224		
7 -	11	37 -	68	67 -	139	97 -	178	127 -	225		
8 -	15	38 -	69	68 -	140	98 -	180	128 -	226		
9 -	21	39 -	70	69 -	141	99 -	183	129 -	227		
10 -	24	40 -	72	70 -	142	100 -	188	130 -	228		
11 -	24	41 -	73	71 -	143	101 -	189	131 -	229		
12 -	25	42 -	74	72 -	144	102 -	190	132 -	230		
13 -	27	43 -	75	73 -	145	103 -	191	133 -	231		
14 -	28	44 -	78	74 -	146	104 -	192	134 -	232		
15 -	30	45 -	91	75 -	147	105 -	193	135 -	233		
16 -	30	46 -	100	76 -	152	106 -	194	136 -	234		
17 -	33	47 -	113	77 -	154	107 -	196	137 -	235		
18 -	38	48 -	115	78 -	158	108 -	199	138 -	236		
19 -	39	49 -	117	79 -	158	109 -	200	139 -	237		
20 -	41	50 -	118	80 -	160	110 -	201	140 -	238		
21 -	42	51 -	119	81 -	161	111 -	202	141 -	239		
22 -	46	52 -	120	82 -	162	112 -	202	142 -	240		
23 -	48	53 -	122	83 -	163	113 -	205	143 -	241		
24 -	49	54 -	123	84 -	163	114 -	210	144 -	242		
25 -	50	55 -	124	85 -	164	115 -	211	145 -	243		
26 -	51	56 -	125	86 -	164	116 -	212	146 -	245		
27 -	56	57 -	127	87 -	165	117 -	213	147 -	246		
28 -	57	58 -	127	88 -	166	118 -	214	148 -	248		
29 -	58	59 -	129	89 -	167	119 -	215	149 -	250		
30 -	59	60 -	130	90 -	168	120 -	216	150 -	250		

TABLE REFERENCES to PAGE NUMBERS

Table No.	Page No.	Table No.	Page No.	Table No.	Page No.	Table No.	Page No.
1 -	2	31 -	87	61 -	121	91 -	190
2 -	2	32 -	87	62 -	126	92 -	197
3 -	5	33 -	88	63 -	128	93 -	197
4 -	13	34 -	88	64 -	128	94 -	198
5 -	14	35 -	88	65 -	137	95 -	198
6 -	15	36 -	89	66 -	151	96 -	203
7 -	16	37 -	91	67 -	153	97 -	204
8 -	17	38 -	91	68 -	157	98 -	206
9 -	22	39 -	93	69 -	157	99 -	222
10 -	23	40 -	93	70 -	157	100 -	225
11 -	23	41 -	93	71 -	159	101 -	227
12 -	29	42 -	96	72 -	159	102 -	229
13 -	31	43 -	97	73 -	161	103 -	231
14 -	35	44 -	99	74 -	165	104 -	233
15 -	36	45 -	101	75 -	166	105 -	235
16 -	37	46 -	103	76 -	168	106 -	237
17 -	41	47 -	103	77 -	172	107 -	239
18 -	44	48 -	104	78 -	172	108 -	241
19 -	45	49 -	105	79 -	173	109 -	243
20 -	45	50 -	105	80 -	177	110 -	257
21 -	54	51 -	106	81 -	179	111 -	257
22 -	55	52 -	107	82 -	179	112 -	258
23 -	60	53 -	108	83 -	181	113 -	258
24 -	65	54 -	108	84 -	182	114 -	259
25 -	65	55 -	109	85 -	183	115 -	261
26 -	82	56 -	110	86 -	184	116 -	262
27 -	83	57 -	114	87 -	186	117 -	262
28 -	84	58 -	114	88 -	187	118 -	263
29 -	85	59 -	116	89 -	189	119 -	267
30 -	87	60 -	119	90 -	190	120 -	267

Index

Accelerated tests 71
Acceptability 53
Acceptability factor 58, 64
Acceptability time 53, 138, 142
Accumulated retention - vitamins 98
Actual freezing rates 36
Additivity 149, 151
Ageing - shelf life 121
Ageing - vitamin retention 95
Air freezing 14, 31, 35, 37
Air thawing 261
Albany 47, 76
Anaerobic packing, see vacuum packaging
Antioxidants 117, 126
Appearance - frozen foods 20, 53, 79
Ascorbic acid 47, 81, 86, 117, 127, 149
Asparagus 35
ATP-Convention 175

B-vitamins 69, 83, 86, 90, 95, 101, 104
Bacon - shelf life 25, 64, 224, 226, 228
Beans - freezing 16
Beans - shelf life 116, 144
Beef - freezing 5, 14, 29, 34
Beef - shelf life 55, 121, 142
Beef - vitamins 83
Berries, see fruits
Best before 254
Biotin, see also B-vitamins 87

Birds Eye   7, 18
Blanching   82, 123, 125
Blast freezing   14, 31, 35, 37
Blast freezing - energy   40
Blast freezing - temperatures   40
Blueberries - freezing   37
Breading   69
Brine freezing, see Ottesen method
Broccoli   69
Broilers - freezing   15, 30, 34, 44
Broilers - shelf life   58, 61, 63, 114, 232

Calcium   84
Cauliflower - shelf life   48, 72
Chicken broilers, see broilers
Chicken parts   59, 69, 113, 234
Chlorophyll   47, 52, 149
Cobolamin, see also B-vitamins   86
Cod   127, 143, 236
Cod fillet - freezing   37
Cold storage flavour   125
Cold stores   165, 210, 211, 215
Colloidal products   152
Colour   47, 48, 49, 149
Compressors   39, 186
Conditioning   121
Congelé   2, 137, 177, 207
Consumer preference   78
Contact plate freezing, see plate freezing
Cook-freeze   94, 108
Cooking   104, 123
Core temperature   40
Cryogenic freezing   14, 40
Cumulative effect, see additivity   149
Cured meats - salt concentration   23

Danish Meat Products Laboratory   62
Danish Meat Research Institute   61
Date marking   253
Deep frozen   208
Denmark - nutrition status   82, 85

Desiccation   151, 204, 206
Display cabinets   28, 182, 188
Display cabinets - temperature   158, 163, 184, 216
Display cabinets - residence time   158, 161, 163, 212
Divider plates   34
Drip   5, 24, 28, 44
Drip method   29
Drip - nutrient loss   81, 102
Drip - thawing rate   31

Electrical stimulation - freezing   14
Electrical stimulation - shelf life   114, 121
End temperature   18, 26, 35, 126
Energy   38, 186, 190, 256
Enzymatic activity   34, 45, 121, 123
Equilibrium temperature   41

Fatty fish   140
Fatty meat   138
Feeding - effect on shelf life   113, 114
FIFO (First in - first out)   174, 186, 193
Fish - freezing   9, 15, 28, 31
Fish - shelf life   122, 125, 127, 139, 143
Fluctuating temperatures, see temperature fluctuations
Fluid bed freezing   35, 37, 125
FND, see also stability time   53
Folates, see also B-vitamins   87
FPD, see also stability time   53
Freezer cabinets, see display cabinets
Freezer chain   153, 156
Freezer chain - time   156, 210
Freezer chain - temperature   156, 215
Freezer storage   44
Freezer storage - nutritive value   97
Freezing - food habits   111
Freezing - reversibility   22
Freezing point   22
Freezing process - nutritive value   81, 96
Freezing process - quality changes   33, 44
Freezing rate   7, 28, 33, 36
Freezing rate - appearance   20, 34

Freezing rate - conclusions   33
Freezing rate - quality   13, 19, 33, 97
Freezing temperatures   40
Freon freezing   15, 35
Frost formation   151, 189, 190, 206
Frozen food - consumption   1, 2
Frozen food - history   1
Frozen meat - further processing   92, 263
Frozen meat - institutional   94
Frozen meat - intake   90
Frozen raw materials   92, 116, 126, 263
Fruits - freezing   15, 31, 35
Fruits - shelf life   49, 71, 126, 134, 140
Fruits - sugar   26, 79, 126, 151

Gefroren   207
Geometric time scale   72, 75
Glass door cabinet   192
Glazing   129
Ground beef   17, 54, 93, 107, 116, 204
Ground pork   54, 204, 266

Haddock   28
Hamburgers - freezing   17, 37
Hamburgers - shelf life   54, 123, 129, 131, 152
Handling   180, 191
Heat radiation   196
Herring   118, 130, 238
Histological studies   7
History - frozen food   1
Home freezers   3, 15, 22, 94, 197, 213, 218, 259
Home transport   181
HQL, see also stability time   53

Ice crystal growth   20, 26
Ice crystals   8, 20
Ice formation   21
Institutional meat supply   94
Internal pressure   7
Intracellular ice crystals   11

Iron in diet	84
Irreversibility - water removal	22, 24
JND, see also stability time	53
Keeping quality, see also PSL	53
Kitchen preparation	104
Labelling	253, 268
Lamb - freezing	14
Lamb - shelf life	26, 114, 126, 132
Leakers	133
Lean fish	139
Lean meat	138
Legislation	43, 136, 148, 253, 265
Lemon sole - freezing	15
Lipids - oxidation	45, 79, 81, 130
Liver - shelf life	55, 120
Loading	175, 180
Logarithmic time scale	47, 73
Meat - B-vitamins	85, 86
Meat - freezing	13, 31, 34
Meat - nutrient content	82, 90, 110
Meat - shelf life	50, 54, 60
Meat balls	37, 51, 189, 230
Meat intake	83, 86
Methodology - TTT tests	47, 60, 78, 149
Microbiology	34, 71, 120, 152
Microwave	261
Mincing	16, 126
Mushrooms - freezing	35
Neutral stability	45, 69
Niacin, see also B-vitamins	86, 98
Night cover	188, 196
Nitrogen freezing	14, 35, 40, 262
Nutrient content of diet	84

Nutrients - daily intake 84
Nutritional aspects 4, 81
Nutritive changes 81
Nutritive changes - ascorbic acid 82, 86
Nutritive changes - B-vitamins 91
Nutritive value 81

Objective tests 47, 79
Orange juice 150
Ottesen method 7, 120

Packaging 113, 128
Packaging date 254
Panel, see also taste panels 77
Pantothenic acid, see also B-vitamins 90
Parsley 133
Peaches 49, 126, 134, 150
Peas 35, 37, 69, 82, 149, 190, 204, 242
Peroxide 79, 114, 130
Pheophytin 47
Pig meat - vitamins 87
Pizza 91
Plate freezing 14, 15, 36, 262
Pork - freezing 13
Pork - shelf life 49, 54, 56, 114, 117, 121, 222
Pork chops 54, 62, 115, 132
Pork sausage - shelf life 54, 74, 119, 124
Port door 176
Potato chips - freezing 37
Poultry - freezing 15, 29, 34
PPP-factors 77, 112, 135
Practical storage life, see PSL
Pre-rigor meat 29, 116
Probability of exposure 220
Process 123
Product 112
Prooxidative effect 118
PSL, see also acceptability time 53, 137
Pudding 152
Pyridoxin (vitamin B_6), see B-vitamins 85

Q10	47, 51, 148
Quality aspects	3
Quality changes - freezer storage	44
Quality loss - calculation of shelf life loss	153
Quick frozen	7, 207
Raspberries	37, 78, 204, 262
Rate of freezing, see freezing rate	
RDA (Recommended Daily Allowance)	83
Ready-to-heat dishes	54, 90, 108, 135, 137, 140, 147
Recrystallization	26, 203
Refreezing	127, 263, 266
Residence times	153, 165
Retail cabinets, see display cabinets	
Retention of nutrients	81, 95, 98
Retinol (vitamin A)	85
Reverse stability	45, 64
Riboflavin (vitamin B_2), see also B-vitamins	85, 98
Sales cabinets, see display cabinets	
Salt - rancidity	117, 120
Salt concentration - cured meats	23
Sausages	54, 57
Scallops - freezing	15
Scoring method - TTT tests	60
Semi-logarithmic diagram	47, 73
Shear force	5
Shelf life	25, 53, 70
Shelf life calculations	153, 209, 244
Shelf life loss	153
Shrimp	15, 128
Sliced bacon, see bacon	
Smoked bacon, see bacon	
Smoking - shelf life	119
Soy protein - shelf life	123
Spinach - freezing	37
Spring rolls	91
Stability time	53, 62, 141
Strawberries	31, 35, 37, 150, 262
Strawberries - shelf life	146, 204, 240
Surgelé	2, 137, 177, 207
Syrup	79, 126

Taste panel   47, 77, 115
Temperature abuse   20, 177, 222, 246
Temperature fluctuations   27, 107, 150, 185, 189, 203
Temperature requirements   137, 207
Temperature rises   176
Tempering   128, 264
Tiefgefroren   207
Time-temperature history   202
Time-temperature indicators   245
Time-temperature integrators   247
Time-temperature surveys   156
Thaw drip, see also drip   44
Thaw rigor   31
Thawed products   265, 268
Thawing   261, 263
Thawing rate   31, 262
Thiamin (vitamin B$_1$), see also B-vitamins   85, 98
Transport   174
Transport - temperatures   176
Triangle test   47, 50
Trouts   127, 128
TTT   47
TTT - methodology   47, 50, 60
TTT-PPP data   137, 220
Turkeys   20, 34, 150, 151

USA - nutrition status   84
Use before   253

Vacuum packaging   50, 65, 72, 118, 124, 128
Vegetables - freezing   15, 35
Vegetables - shelf life   47, 125, 139
Vertical cabinets   191
Vitamins in frozen foods   69, 81
Volatile flavour substances   134

Warmholding   108
Water - frozen out   21, 22
Water binding capacity   16, 45
Wholesale storage, see cold stores   174, 211, 215

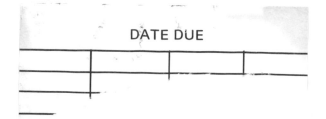

DATE DUE